[illegible]
[illegible]
Toronto, ON M8V 1K8

Locative Media

"Wilken and Goggin's edited collection is a valuable addition to the literature on locative media. By combining an international perspective with a wide range of topics, this book helps further our understanding of the potential impacts of these new technologies. One of the true strengths of this book is how it combines cultural, political, and economic areas of focus, and it will be a valuable resource for anyone studying the emergence of new mobile technologies."—*Jordan Frith, University of North Texas, USA*

Not only is locative media one of the fastest growing areas in digital technology, but questions of location and location awareness are increasingly central to our contemporary engagements with online and mobile media and indeed with media and culture generally. This volume is a comprehensive account of the various location-based technologies, services, applications, and cultures, as media, with an aim to identify, inventory, explore, and critique their cultural, economic, political, social, and policy dimensions internationally. In particular, the collection is organized around the perception that the growth of locative media gives rise to a number of crucial questions concerning the areas of culture, economy, and policy.

Rowan Wilken is Senior Lecturer in the Faculty of Health, Arts and Design, and a postdoctoral research fellow in the Swinburne Institute for Social Research, at Swinburne University of Technology, Melbourne, Australia.

Gerard Goggin is Professor and Chair of the Media and Communications Department at the University of Sydney, Australia.

Routledge Studies in New Media and Cyberculture

1 **Cyberpop**
Digital Lifestyles and Commodity Culture
Sidney Eve Matrix

2 **The Internet in China**
Cyberspace and Civil Society
Zixue Tai

3 **Racing Cyberculture**
Minoritarian Art and Cultural Politics on the Internet
Christopher L. McGahan

4 **Decoding Liberation**
The Promise of Free and Open Source Software
Samir Chopra and Scott D. Dexter

5 **Gaming Cultures and Place in Asia-Pacific**
Edited by Larissa Hjorth and Dean Chan

6 **Virtual English**
Queer Internets and Digital Creolization
Jillana B. Enteen

7 **Disability and New Media**
Katie Ellis and Mike Kent

8 **Creating Second Lives**
Community, Identity and Spatiality as Constructions of the Virtual
Edited by Astrid Ensslin and Eben Muse

9 **Mobile Technology and Place**
Edited by Gerard Goggin and Rowan Wilken

10 **Wordplay and the Discourse of Video Games**
Analyzing Words, Design, and Play
Christopher A. Paul

11 **Latin American Identity in Online Cultural Production**
Claire Taylor and Thea Pitman

12 **Mobile Media Practices, Presence and Politics**
The Challenge of Being Seamlessly Mobile
Edited by Kathleen M. Cumiskey and Larissa Hjorth

13 **The Public Space of Social Media**
Connected Cultures of the Network Society
Thérèse F. Tierney

14 **Researching Virtual Worlds**
Methodologies for Studying Emergent Practices
Edited by Ursula Plesner and Louise Phillips

15 **Digital Gaming Re-imagines the Middle Ages**
Edited by Daniel T. Kline

16 Social Media, Social Genres
Making Sense of the Ordinary
Stine Lomborg

17 The Culture of Digital Fighting Games
Performances and Practice
Todd Harper

18 Cyberactivism on the Participatory Web
Edited by Martha McCaughey

19 Policy and Marketing Strategies for Digital Media
Edited by Yu-li Liu and Robert G. Picard

20 Place and Politics in Latin American Digital Culture
Location and Latin American Net Art
Claire Taylor

21 Online Games, Social Narratives
Esther MacCallum-Stewart

22 Locative Media
Edited by Rowan Wilken and Gerard Goggin

Locative Media

Edited by Rowan Wilken
and Gerard Goggin

HUMBER LIBRARIES LAKESHORE CAMPUS
3199 Lakeshore Blvd West
TORONTO, ON. M8V 1K8

First published 2015
by Routledge
711 Third Avenue, New York, NY 10017

and by Routledge
2 Park Square, Milton Park, Abingdon, Oxon OX14 4RN

Routledge is an imprint of the Taylor & Francis Group, an informa business

© 2015 Taylor & Francis

The right of the editors to be identified as the author of the editorial material, and of the authors for their individual chapters, has been asserted in accordance with sections 77 and 78 of the Copyright, Designs and Patents Act 1988.

All rights reserved. No part of this book may be reprinted or reproduced or utilized in any form or by any electronic, mechanical, or other means, now known or hereafter invented, including photocopying and recording, or in any information storage or retrieval system, without permission in writing from the publishers.

Trademark Notice: Product or corporate names may be trademarks or registered trademarks, and are used only for identification and explanation without intent to infringe.

Library of Congress Cataloging-in-Publication Data

Locative media / edited by Rowan Wilken and Gerard Goggin.
pages cm — (Routledge studies in new media and cyberculture ; 22)
1. Location-based services—Social aspects. 2. Mobile computing—Social aspects. I. Wilken, Rowan. II. Goggin, Gerard, 1964–
HM851.L6535 2015
004—dc23
2014010248

ISBN: 978-0-415-70708-4 (hbk)
ISBN: 978-1-315-88703-6 (ebk)

Typeset in Sabon
by Apex CoVantage, LLC

Printed and bound in the United States of America by Publishers Graphics, LLC on sustainably sourced paper.

Rowan Wilken:
For Karen.

Gerard Goggin:
For Jacqueline *otra vez.*

Contents

List of Figures xiii
Acknowledgments xv

1 Locative Media—Definitions, Histories, Theories 1
ROWAN WILKEN AND GERARD GOGGIN

PART 1
Practices, Publics, Spaces

2 Intimate Cartographies of the Visual: Camera Phones, Locative Media, and Intimacy in Kakao 23
LARISSA HJORTH

3 Mobile Communication Technologies and Spatial Perception: Mapping London 39
DIDEM ÖZKUL

4 The Social Media Life of Public Spaces: Reading Places Through the Lens of Geotagged Data 52
RAZ SCHWARTZ AND NADAV HOCHMAN

5 Locative Praxis: Transborder Poetics and Activist Potentials of Experimental Locative Media 66
ANDREA ZEFFIRO

PART 2
Geography, Code, Representation

6 Map Interfaces and the Production of Locative Media Space 83
JASON FARMAN

7 Locating Media, Performing Spatiality: A Nonrepresentational Approach to Locative Media 94
FEDERICA TIMETO

8 The Cluster Diagram: A Topological Analysis of Locative Networking 107
CARLOS BARRENECHE

PART 3
Information, Privacy, Policy

9 Evolving Concepts of Personal Privacy: Locative Media in Online Mobile Spaces 121
TIMOTHY DWYER

10 Google Glass and Australian Privacy Law: Regulating the Future of Locative Media 136
JAMES MEESE

11 Locative Media, Privacy, and State Surveillance in Mexico: The Case of the Geolocalization Law 148
GERARD GOGGIN AND CÉSAR ALBARRÁN-TORRES

12 Seeking Transparency in Locative Media 162
TAMA LEAVER AND CLARE LLOYD

PART 4
Economies, Networks, Logistics

13 Locating Foursquare: The Political Economics of Mobile Social Software 177
ROWAN WILKEN AND PETER BAYLISS

14 Becoming Drones: Smartphone Probes and Distributed Sensing 193
MARK ANDREJEVIC

15 Locative Media as Logistical Media: Situating Infrastructure and the Governance of Labor in Supply-Chain Capitalism 208
NED ROSSITER

16 Locative Esthetics and the Actor–Network 224
MICHAEL DIETER

Contributors 237
Index 243

Figures

2.1 Sol's KakaoStory, image 1. 31
2.2 Sol's KakaoStory, image 2. 32
3.1 (a) Mary's sketch map of London. (b) "Design Museum and Tate are like two checkpoints for me to see what is where in the East" (Mary). 42
3.2 (a) Marianne's map of London. (b) Marianne's running route along the Thames. 48
4.1 Geographical plotting of publicly shared geotagged Instagram photos from November 20, 2012–April 15, 2013. Left: Union Square—14,593 photos, Center: Bryant Park—18,352 photos, Right: Madison Square Park—14,989 photos. 57
4.2 (a) Hourly time series of publicly shared geotagged Instagram photos November 20, 2012–April 15, 2013. Left: Union Square—14,593 photos, Center: Bryant Park—18,352 photos, Right: Madison Square Park—14,989 photos. (b) Daily time series plotting of publicly shared geotagged Instagram photos November 20, 2012–April 15, 2013 (number of photos per park as before). 58
4.3 Word cloud plotting of Yelp Review: (a) from Union Square—226 reviews; (b) from Bryant park—462 reviews; (c) from Madison Square Park—221 reviews. 60
4.4 Phrase Net visualization displays networks of related words of Yelp Review: (a) from Union Square—226 reviews; (b) from Bryant park—462 reviews; (c) from Madison Square Park—221 reviews. 61
5.1 Ricardo Dominguez during a research trip to Calexico, California, a town on the U.S.–Mexico border, in 2008 (photograph by Brett Stalbaum). 67
5.2 The Transborder Immigrant Tool billboard for Galeria de la Raza in San Francisco (photograph by Ricardo Dominguez). 71

5.3 The Transborder Immigrant Tool in operation—with a screenshot from a Nokia e71 cell phone—directing a user to a Water Station Inc. water cache in the Anza Borrego Desert, located within the Colorado Desert of southern California (photograph by Brett Stalbaum). 74
5.4 Brett Stalbaum shortly after he and Ricardo Dominguez had been pulled over by the border patrol on Highway 98 near Calexico; Stalbaum had just finished talking to the officers (photograph by Ricardo Dominguez). 76
7.1 Julian Oliver, *Border Bumping* (2012–2013) (© Crystelle Vu, 2013). 97
7.2 Julian Oliver, *Border Bumping* (2012–2013) (© Julian Oliver, 2013). 98
8.1 Nearest-neighbor classification diagram (source: "*k*-nearest neighbor algorithm," Wikipedia, May 28, 2007; upload by Antti Ajanki). 110
8.2 Nearest-neighbor network topology (source: "Nearest Neighbor Networks," Wolfram Demonstrations Project, accessed December 10, 2013. www.demonstrations.wolfram.com/NearestNeighborNetworks/). 112
9.1 Examples of mobile applications functionality 124
9.2 Mobile application information flows 128

Acknowledgments

Edited collections are, by definition, a collaborative enterprise, and many people contribute to their successful completion. At Swinburne University of Technology, we would like to thank Julian Thomas, Anthony McCosker, Jenny Kennedy, James Meese, Peter Bayliss, and the administrative staff of the Swinburne Institute for Social Research. At the University of Sydney, we would like to thank Gerard's colleagues in the Department of Media and Communications. Further afield, we would also like to thank Lee Humphreys.

A very special mention and debt of gratitude goes to Emily van der Nagel for all her hard work, done with such efficiency and good humor, in helping pull this collection together.

We also wish to express our gratitude to Felisa Salvago-Keyes at Routledge for supporting and helping to realize this collection. And, of course, to all the excellent contributors to this collection for their interest and faith in this project, for their willingness to contribute their work, and for the patience and good humor with which they have responded to our requests, promptings, and questions, often within very tight timeframes.

This book is an output from a program of research under an Australian Research Council (ARC) Early Career Researcher Award—DE120102114, "The Cultural Economy of Locative Media," funded 2012–2014. We gratefully acknowledge the ARC's financial support.

Finally, we also wish to thank our families. Rowan would like to thank Karen, Lazarus, Maxim, and Sunday, for their love, support, and encouragement. Gerard thanks Jacqueline, Liam, and Bianca for their love and forebearance.

Rowan Wilken, Swinburne University of Technology
Gerard Goggin, The University of Sydney
June 2014

1 Locative Media—Definitions, Histories, Theories

Rowan Wilken and Gerard Goggin

INTRODUCTION

In Seoul, a young woman uses the location-based social networking service Kakao in order to tag, upload, and share a photo just taken of their catch-up with friends. In Helsinki, a family plays *Angry Birds* together, as the app gathers information on their location via the smartphone and its location technologies and sensors. In Chongqing, middle-class café habitués are sharing contacts, messages, and photos via the networking application Jiepang (which now includes the trendy mobile app WeChat), popularized in cities like Shanghai. In rural United States, a child calls 911 emergency services for help, and the ambulance is dispatched using the location information available via their absent parents' phone.

Every day, tens of millions of mobile users navigate and way-find using mobile maps that pinpoint their location. Around the world, men flirt, connect, and hook up with other men by using Grindr and other apps to scan for who is free and available in their vicinity. Crossing the border from Mexico to the United States, immigrants use GPS to find food, water, and help, while their movements trigger embedded sensors in the landscape, alerting authorities to their presence. Elsewhere in Mexico, law enforcement agents use geolocation data on a suspect's whereabouts, provided by a mobile phone company.

In Africa, a mobile developer at a trade fair in Johannesburg pitches the virtues of locative mobile media, hoping to build on pioneering efforts in user-generated news and citizen journalism such as Ushahidi. Meanwhile, all across the African continent, users continue to routinely use text messaging on cheap mobiles to tell friends, family, and customers where they are. In New South Wales, Australia, a resident files a privacy complaint after he wakes up to the company of a drone hovering outside his bedroom window. In San Francisco, a tech enthusiast-cum-writer is assaulted in a bar for the offense of sporting her new Google Glass.

These vignettes indicate some of the growing range of technologies, features, uses, practices, meanings, emotions, and possibilities that are associated with locative media. Broadly speaking, locative media involves the use of information, data, sounds, and images about a location. In reality, as we shall discuss in this chapter and as will be borne out in this book, the definitions of both "locative" and "media" turn out to be much more complex, fabulous, prosaic, frustrating and disappointing than they might seem.

If the locative aspect is tricky to pin down, then, in the second decade of the twenty-first century, the media part of the couplet is expanding almost beyond comprehension. So locative media involve not just global positioning satellites (GPS), cellular mobile phones, location-based services (LBS), and apps. Locative media have gone well beyond mobile social software, social networking applications, and so-called check-in applications from Dodgeball and Foursquare, through Facebook Places, Twitter, and Weibo. It is turning out that locative media are the harbinger of the emergent media of our time, from big data to drones, from the Internet of Things to logistics, all with their urgent cultural, social, and political implications.

Hence this volume aims to provide a state-of-the-art snapshot and analysis of contemporary locative media, along with the ideas, technologies, practices, contexts, and power relations shaping them and the research underway that seeks to illuminate them. Accordingly, in this opening chapter, we provide an introduction to locative media. In the first two sections, we discuss the cardinal concept of location and chart the histories and development of locative media. In the third section, we provide an overview of the key theoretical tributaries feeding research into locative media, allowing us also to situate the various contributions to this volume.

THE CONCEPT OF LOCATION

The concept of location (like the related concepts of place and space) has developed according to a diverse and complicated set of etymological trajectories that include legal use (with location understood as the action of letting for hire) and grammatical use (where it refers to a particular case form).[1] It also refers to land settlement practices, as well as to processes of emplacement and "the action of discovering, or the ability to discover or determine, the position of a person or thing."[2] It is these last two senses that constitute the general understanding of the term and that most inform how the term is employed in relation to mobile media technologies.

In addition to its rich etymological history, location is also a foundational concept in classical geography, where it is considered to act as a unifying concept.[3] As Fred Lukermann explains, for classical geographers, "the essence of all the methods is *location*"; it is the umbrella term that unifies geography, chorography, and topography.[4] For instance, in the philosophy of Aristotle, Lukermann writes that "place or space description is simply an analysis of relative location."[5] For Strabo, "what seemed in the past to be either divergent self-contained trends are brought together into comprehension around the central thesis of location."[6] Location is also a concept, Lukermann asserts, that carries important general significance. From the Ancient Greek poets onward, he writes, "how to describe 'where something is' becomes idiomatic in Western culture."[7]

Nevertheless, location has since come to be understood as a subsidiary concern to the more encompassing concept of place. This is a perspective that is captured in Ed Relph's description of place as "location plus everything that occupies that location seen as an integrated and meaningful phenomenon."[8] More recently, however—and commensurate with the rise of location-enabled mobile communications technologies—location is viewed as having taken on increased (or renewed) conceptual importance in its own right. Adriana de Souza e Silva and Jordan Frith encapsulate this renewal of interest in the concept of location as follows:

> The popularization of location-aware mobile technologies not only highlights the importance of location, but also forces us to re-think how location has been traditionally conceptualized. Locations are still defined by fixed geographical coordinates, but they now acquire dynamic meaning as a consequence of the constantly changing location-based information that is attached to them.[9]

Thus, they argue, where locations were once seen as "places deprived of meaning" (or perhaps, whose meaning was dependent on other concepts and phenomena), they can now be seen as taking on "complex, multifaceted identities that expand and shift according to the information ascribed to them."[10]

The efficacy of the term "locative media" in describing recent developments in mobile media technologies, as Minna Tarkka notes, is also a product of the meanings carried by it.[11] For instance, media arts theorist Karlis Kalnins, the person generally credited with coining the term "locative media" (broadly defined as media of communication that are functionally bound to a location), is understood to have been drawn to the word "locative" based on his knowledge of "languages such as Latvian and Finnish with their several locative cases corresponding roughly to the preposition 'in,' 'at,' or 'by,' and indicating a final location of action or a time of the action."[12] This etymological preference on Kalnins' part is more than mere semantics; it is deliberate insofar as, for him, it strategically repositions media arts practice by shifting the emphasis off the site of action (actual places or locations) and onto the agency and actions of subjects and onto the temporal dimensions of these actions. Our perspective is slightly different. Although continuing to use the term "locative media," we acknowledge the significance of Kalnins' linguistic preference for "locative" over "location;" our preference is that the two—actions *and* the sites in which these temporally based actions occur—be kept in a productive tension that also accounts for the various technological (cellphone towers, radio signals, handsets, Wi-Fi, etc.) and other infrastructures (not the least of these being corporate arrangements) that mediate our locationally situated technosocial interactions.[13]

HISTORIES AND DEVELOPMENT OF LOCATIVE MEDIA

Location technologies have experienced a relatively long and complex incubation. The satellite-based global positioning system (GPS) commenced life as a military technology before finding its way into wider commercial and consumer uses (not least of which is being used in mobile phones, along with triangulation of cellular networks for such services as enhanced emergency calling). Location-based services for cellular mobile networks and devices were the subject of much experiment and anticipation in the 1990s. Mobile social networking applications first emerged in the 1990s, with the celebrated Lovegety gadget in Japan, and pioneering efforts such as Dodgeball in North America. Technologies predicated on location were also pieced together through telecommunications, the Internet, and web-based friendship, dating, and hooking-up services and sites such as Gaydar.

The early 2000s witnessed a wave of location-based experimentation with location and mobile devices across art, urban design, ubiquitous and pervasive computing, and strands of gaming cultures. These experiments included locative art, performances, activist interventions, location-aware fiction, location-based games (famously those of Blast Theory), annotation, and storytelling. As mobile phones developed into full-fledged media devices, various affordances led to new kinds of sociotechnical marshaling of location. The ubiquity of camera phones allowed innovative visual and textual instantiations and representations of place. Cross-platform game developments increasingly relied on locative media as a key part of integrated, transmedia forms. Music and sound moved to the foreground of media, which were imaginatively yoked to location.

In this book, "locative media" is the term that is used to capture this diverse array of location-aware technologies and practices. The term "locative media" (that is, media of communication that are functionally bound to a location) is preferred for the simple reason that it is economical and expansive but also precise. That is to say, it captures a lot in two words while retaining a sense of the term's very particular history, which is anchored in the field of new media arts.[14]

Whatever the precise origins of the term, it is fair to say that the field of new media arts has been at the vanguard of exploring both the creative possibilities and the critical implications of locative media, and it is where the bulk of the literature on locative media to date is found. Here, important work has been done, to cite just two examples, in exploring how location-based services can generate new potentialities for facilitating the forms of social appropriation, citizenship, and (experimental) sociability[15] and in examining the "particularities, tensions and conflicts" associated with urban space.[16]

Outside of media arts, significant work has been done on locative media at the intersection of research into mobile technologies, geography (particularly the subfield of media geography),[17] and urban space and place. Taken up in this body of work are myriad considerations, which range across (to name only a few) analysis of how locative technologies mediate the relationship between technology use and physical/digital spaces;[18] exploration of the representation of space and spatial practice through locative media;[19] and concern for what might be described as questions of power and the politics of location and locatability.[20] What these examples evidence, in short, is a flowering of detailed, wider, interdisciplinary scholarship on and around locative media.

Nowadays, not only do location-based services "comprise the fastest growing sector in web technology businesses,"[21] questions of location and location awareness are increasingly central to our contemporary engagements with the Internet and mobile media. Indeed, as Eric Gordon and Adriana de Souza e Silva suggest, "unlocated information will cease to be the norm,"[22] and location will become a "near universal search string for the world's data";[23] or, as Malcolm McCullough puts it, information "is now coming to you . . . wherever you are" and "is increasingly *about* where you are."[24] Toward the end of the first decade of the twenty-first century, three major developments put locative media squarely at the center of contemporary cultural and social dynamics and fueled this interest in location-tagged data.

First, new kinds of locative media emerged through the "geoweb"—the combination of the Internet with mapping, place-making, and locational technologies. Since Google's embrace of geolocation services in 2005—with the fascination attracted by Google Earth and Google Maps—mainstream interest in and uptake of locative media services flourished. As Gordon and de Souza e Silva explain in relation to mapping technologies:

> For many decades, the geolocation industry was focused on developing high-end geographic information system (GIS) software for market and social research, as well as military purposes. But when Google Maps launched in February 2005, and its application programming interface (API) was made available to the public just a few months later, the specialized domain of GIS programmers became the domain of everyday users.[25]

Such Internet-based locative media increasingly coincided with the widespread diffusion of mobile phone, mobile broadband, wireless Internet, and portable, networked media technologies. Consumers are now well accustomed to using sat nav devices in their cars or while walking, Google Maps on desktop and laptop computers and mobile devices, and geoweb, geotagging, and other mapping applications from all manner of places, as well as various apps on iPhones and smartphones that use location-aware technologies.

Second, with the phenomenal growth of smartphones following the launch of Apple's iPhone and Google's Android platforms in 2007–2008, the mobile Internet firmly took hold. According to Gerard Goggin, the arrival of the smartphone—the "iPhone moment," as he refers to it—was significant in that it "galvanized users, developers, industry, policy makers and a range of publics alike, to articulate their concerns and desires regarding mobile media"[26] and facilitated the rapid wider take-up of locative media services. The rise of the smartphone accelerated the trend toward the crossover between Wi-Fi (wireless internet) and cellular mobile networks and devices. The handsets, applications, and users now switched with ease among networks. This capability was also reflected in the evolution of the network architectures and infrastructures, as telecommunications and mobile networks—especially next-generation networks and 4G and 5G mobile networks—merged with Internet protocol and data networks.

Furthermore, as inadvertently revealed, smartphones gather unprecedented amounts of longitudinal data on their users' locations—data that can support new kinds of tailored retail and consumer services, lifestyle profiling and mapping, and surveillance, all with considerable privacy and social implications. Such mobile media built on the success of user-generated content and social networking systems (Cyworld, Mixi, Flickr, YouTube, QQ, Renren) brought the locational aspects of these systems to the fore, especially with extensions such as Facebook Places, iPhoto tagging, and so on. The arrival of apps on smartphones—supported by Apple's apps store, Google market, Windows' app store, and Nokia's shared apps—was also fueled by the incorporation of locational capacities into this new wave of mobile computing and software.

Third, an enormous growth in personal, private, and machine-based information and processing has been associated with a wide range of consumer and enterprise technologies and networks, which adds significantly to mass personalized user, device, and network data concerning location. The much vaunted Internet of Things, associated with RFID chips in products especially,[27] is one aspect of this. Another development is the growth in technologies of sensors that are now connected to networks[28] and that generate vast, little understood, and poorly conceptualized troves of data.[29] Ubiquitous computing is a relatively long-standing movement and an important precursor to mobile computing that now overlaps with locative media in particular.[30] Also worth noting is the smart cities movement, a confluence of such trends in its own right that provides another important way to appropriate, deploy, and frame locative media.[31]

These are but three developments that have had a profound impact on fostering the democratization of, and the opening up of access to, geolocation services and associated infrastructure. As a result, not only are locative media among the fastest growing areas in digital technology, questions of location and location awareness are central to our contemporary engagements with online, mobile media, and media and culture in general.

THEORETICAL TRIBUTARIES FEEDING LOCATIVE MEDIA RESEARCH

In addition to these three specific sociotechnical developments, contemporary academic research on locative media, especially as it is represented in this volume, is fed from diverse theoretical tributaries: interdisciplinary examinations of everyday life; critical reevaluations of representation, software and critical code studies, materiality, and the status of objects; and political economy, politics, and policy. Each of these will be addressed in turn and linked to the chapters that combine to form this collection.

Everyday Practice and the Problem of Representation

One of the truisms about everyday life—rendered most famously by Maurice Blanchot[32]—is that, in its very immediacy, in its very ordinariness, the everyday evades capture and analysis—a condition that, at least in part, is a result of what Michael Foley refers to as "the anaesthetising effect of familiarity and habit."[33] The result of which, Henri Lefebvre writes, is that everyday life "takes on the distance and remoteness and familiar strangeness of a dream."[34] And yet it is precisely this elusiveness and familiarity that points to its significance and suggests that these everyday routines and actions add up to more than the sum of their parts: "the everyday is the accumulation of 'small things' that constitute a more expansive but hard to register 'big thing.'"[35] This dynamic between detail and big picture, as well as the significance of both, is nicely captured by Lefebvre when he writes that "the everyday is . . . the most universal and the most unique condition, the most social and the most individuated, the most obvious and the best hidden."[36]

In key respects, critical accounts of the everyday convey, at their heart, critiques of the emergence, continuing development, and ongoing impacts of modernity—and especially of what Highmore terms "everyday modernity" (modernity as experienced as lived practice).[37] A crucial consideration within contemporary studies of "everyday modernity" is the critique of the "interactions between everyday life and technologically driven communication practices."[38] This has long been of interest to mass media researchers and is also of enduring interest to mobile communications scholars. What the second of these forms of media adds to this examination of the everyday life and communication practices is the dynamic aspect of mobility and how movement through space and time impacts and shapes our everyday engagements with portable media and communications devices—and vice versa. Moreover, as we have argued elsewhere, mobile media research that is informed by (post-)phenomenological theory has paid close attention to "the question of how to account for fluctuating levels of attentiveness" and the "specific and complex nature of user engagement with mobile handheld screen devices," which involve quite particular sets of body–technology relations.[39]

In the present collection, Larissa Hjorth's chapter builds on this larger body of mobile-related scholarship and makes an important (internationalizing) contribution to the study of mobile-mediated everyday practice. In her examination of female use of Kakao, a free messaging and social media platform built specifically for mobiles in South Korea, Hjorth explores how camera phone image taking and sharing "can be viewed as part of a broader process of 'emplacement' within everyday movements," practices that involve various forms of "ambient play." Hjorth's larger argument is that this gendered camera phone use contributes in vital ways to the construction of "intimate cartographies," which she defines as forms of "sociality and intimacy" that are "overlaid on the geographic and spatial as part of everyday movements."

But what of "everyday movements" that are not characterized by (or as) "relentlessly routinized" practices,[40] however interesting and "familiarly strange" these may be, but that are rather infused with heightened risk and the omnipresent threat of danger? What role might locative media technologies play in ameliorating these effects? These are questions taken up by Andrea Zeffiro in Chapter 5, which makes an important contribution to a growing body of work on "activist tech." Zeffiro develops the concept of "locative praxis," defined as "a conceptual framework for understanding the ways in which experimental locative media might engage in political and cultural activism and dissent." The focus of her chapter is on an examination of the highly controversial Transborder Immigrant Tool (a project that aims to assist Mexican immigrants crossing the U.S. border), which Zeffiro reads as a form of "locative praxis" that "articulates a political dimension of locative media through the dialectic of practice and reflection, at the intersection of social action and intent." One important role of locative praxis, Zeffiro argues, is in furthering "understanding of the spatial and socio-political dimensions of a particular space/place," such as the U.S.–Mexican border.

Returning to scholarship on everyday life, Ben Highmore argues that in thinking about how we might begin to engage critically with the everyday as lived practice, questions of representation and aesthetics are also crucial. "The necessity of fashioning new forms (or tools) for apprehending new kinds of experiences (new 'realities')," he writes, "might be seen as the general impetus and problematic attendant on theorizing daily life."[41] For Raz Schwartz and Nadav Hochman (Chapter 4), location-based social media systems both complicate such considerations *and* provide a productive means of attempting to grasp the complicated individual and sociospatial practices of everyday locative media users. In their contribution to this volume, they write: "By applying computational tools to data sets from various location-based social networks, we are able to identify special characteristics and produce a computational analysis of these places," thus leading to potentially more nuanced understandings of them. In addition to depicting "what 'the social media image' of a public place is," they also reflect on some of the challenges and limitations of social data-driven locative media research.

Questions of representation and everyday practice are also the focus of Jason Farman's and Didem Özkul's chapters, both of which examine location-based media and map use. Farman opens with an account of the controversy that followed the introduction of Apple Maps on iOS. This was significant, Farman suggests, because it caused "a change in the entire experience of using an iPhone as a means to orient oneself in space and understand that space in a context-specific way." Farman uses this case as a point of entry for thinking through the links between embodiment, practice, and representation. Drawing on the locative media art of Esther Polak, Farman argues that her projects draw attention to "the tensions between our assumptions about our locations and our role in the production of space," prompting us to think critically about the interrelationships between locative media use, corporate interests and stakes, and subjectivity.

Özkul, meanwhile, explores how "locational information use on mobile communication devices changes spatial practices and navigation in London in relation to locations and places." Her interest is in people's personal experiences of urban navigation and how mobile maps can play a valuable role in these individual experiences of everyday practices of spatial movement, navigation, and exploration.

The very issues that Highmore raises, as previously explained, concerning how we "represent" everyday practice have led geographers such as Nigel Thrift (and others) to argue for critical approaches—commonly gathered under the term "nonrepresentational theory"—that are attentive (in Thrift's words) to "the geography of what happens" and that attempt to describe (including via experimental means) "the bare bones of actual occasions"[42] (or, to use Deleuzean terminology, of "events")[43] and to better account for "our self-evidently more-than-human, more-than-textual, multisensual worlds."[44] As Hayden Lorimer explains, the focus of nonrepresentational theories

> falls on how life takes shape and gains expression in shared experiences, everyday routines, fleeting encounters, embodied movements, precognitive triggers, practical skills, affective intensities, enduring urges, unexceptional interactions and sensuous dispositions . . . which escape from the established academic habit of striving to uncover meanings and values that apparently await our discovery, interpretation, judgment and ultimate representation.[45]

Nonrepresentational theories provide the anchor for Federica Timeto's chapter. Timeto argues that we need to question "the representational paradigm" "in order to prioritize the *processes* and *relations* that characterize locative media." Timeto draws from nonrepresentational theories in order to "try and reconnect bodies and geophysical worlds at the interface between matter and information." The value of these theoretical approaches, she argues, is that they can lead to a "rearticulation of space and representation" so that space "comes to be seen as a heterogeneous domain of relations that require continuous engagement" and representation "becomes a situated practice that does not depict reality from afar but which contributes to its construction and transduction from within." In his chapter, Michael Dieter takes a slightly different, although not altogether unrelated, line of argument to that developed by Timeto. Dieter, who draws from the recent work of Bruno Latour rather than Thrift and other nonrepresentational theorists, is interested in the question of "how locative media art might be theorized today as a form of critical technical practice," one that is "invested in modes of problem-creation and discovery within the complexities of net localities and common atmospheres." Dieter contends that ANT-style accounts of mobile technologies would do well to think about esthetics "in relation to the contingencies of inequality in the habitual reproduction of locative media"—that is, to examine esthetics in relation to problems, not in terms of solutions. Building this argument via a detailed engagement with the creative projects of James Bridle, Julian Oliver, and Philip Ronnenberg, Dieter traces the ways that these artists' works "examine how location-aware devices support processes of explication and network curtailment in the production of new cartographies."

Materiality, Space and Software, and the Force of Objects

Building on established work in anthropology and material culture studies,[46] there has, in recent times, been a renewal of interest within media and communications and related disciplines in critical examinations that explore the materiality of technological and other objects, both as "things" and in overtly relational contexts, in terms of their origins,[47] how they shape what we do,[48] and how they connect and interoperate.[49]

One influential strand of work that has emerged from this reinvigorated theoretical interest in materiality and that informs locative media scholarship comes under the rubric of "new materialism."[50] According to one account, "new materialism" advances the claim that "foregrounding material factors and reconfiguring our very understanding of matter are prerequisites for any plausible account of coexistence and its conditions in the twenty-first century."[51] Influenced by the earlier work of Foucault and Deleuze (among other reference points), "new materialism" is said to give explicit focus to "the emergence of pressing ethical and political concerns" that arise from a range of scientific, technological, and social developments, including the "saturation of our intimate and physical lives by digital, wireless, and virtual technologies."[52]

As Jeremy Packer and Stephen Crofts Wiley point out, this has led to recent work that focuses on what they term the "materiality of communication," that is, "communication infrastructure, transportation and mobility, mobile technologies, and the production of urban, regional, and translocal spaces."[53] Within existing locative media scholarship, for instance, Adriana de Souza e Silva and Jordan Frith give consideration to the ways that location-based mobile technologies and practices alter users' experiences of and engagements with urban space and place by allowing users "to interact with the digital information embedded in that space" and thereby "control how they manage their interactions with nearby people and information."[54]

What de Souza e Silva and Frith's work points toward is the importance of accounting for the manifold ways by which mobile-mediated experiences, engagements, and interactions with urban locations are inevitably "tuned"[55] by software. Rob Kitchin and Martin Dodge refer to this as "code/space," the situation "when software and the spatiality of everyday life become mutually constituted, that is, produced through one another."[56] For Martin Dodge, Rob Kitchin, and Matthew Zook, the implications of "code/space" are far-reaching. Of greatest concern to them is the "growing calculative role of code . . . with its ability to render all kinds of spaces 'machine readable.' "[57] One implication of this, they argue, is that "these extensive and readable spaces can then by interpreted by code which makes decisions automatically including socially significant ones, in terms of access, control, and anticipatory governance."[58]

These are issues that are taken up and explored in a number of contributions to the present collection. For instance, building on earlier work of his that examines how places are ordered and sorted,[59] Carlos Barreneche, in Chapter 8, explores the "geodemographic spatial rationality" that underpins and drives commercial locative media services. Unlike traditional forms of geodemographic ordering, where the basic unit of measure is the household, in locative media systems, such as Foursquare, the key unit of measure is the place (or venue) as characterized via points-of-interest databases. These databases, Barreneche argues, assume that "social identity corresponds to spatial mobility," data on which is tracked and fed into corporate software systems. Central to this process, Barreneche concludes, are clustering classification techniques that ultimately suggest "enforced" homophilic tendencies that govern how "we sort and navigate the world."

Given the core role that mobile devices play in how we move through and sort the world, Mark Andrejevic argues, in Chapter 14, that it makes sense to conceive of location-enabled portable devices as "drone-like" insofar as they silently gather information that "track[s] an 'activity stream';" that is to say, locative media devices work to an "emerging logic of portable, always-on, distributed, ubiquitous, and automated information capture." Andrejevic builds this argument via an examination of two applications—one commercial and one U.S. government–created—that act as portable drone devices extracting environmental sensor data. The chapter concludes, in Andrejevic's words, "with a discussion of what might be described as 'drone theory:' the fascination with posthuman forms of information processing in an era of automated data collection and response."

Ned Rossiter also explores the implications of location within large-scale software systems. However, rather than focusing on the richness of data that is associated with the registering of precise user location (as is the concern of Barreneche and Andrejevic), Rossiter takes a different angle of approach. In Chapter 15, he substitutes *distribution* for nearness in his examination of the logistical software developed by the German-based company, SAP. By rethinking locative media as logistical media ("locative media are media of logistics"), Rossiter is interested in understanding the complicated interconnections among location, software, and infrastructure that "coordinate and control the movement of people, finance, and things" and are part and parcel of "supply chain capitalism." Rossiter concludes his chapter by proposing the development of "logistical media theory." Informed by "new materialism" approaches, "logistical media theory," as he conceives of it, serves as a "framing device" for future research and combines (1) a concern for large-scale logistics infrastructure (such as ports) with (2) critical examinations of digital communications technologies and software and (3) detailed accounts of software and logistics industry labor practices.

One consequence of our complicated interactions with communications technologies and associated infrastructure, as suggested by Rossiter's chapter, is the growing number of calls "to come up with elaborated ways to understand how perception, action, politics, [and] meanings . . . are embedded not only in human and animal bodies, but also in much more ephemeral, but [just] as real, *things*—even non solid things."[60] This understanding of human–object relationships and their construction is a key aspect of established accounts of actor–network theory (ANT).[61] However, a focus on the significance of objects is even more overt in the work of those media theorists and philosophers who argue in support of an "object-oriented ontology." Object-oriented ontology (or triple O, as it sometimes called) is an umbrella concept or philosophical orientation (incorporating related approaches, including "flat ontology," "tiny ontology," and "alien phenomenology") that seeks to develop a thoroughly democratic approach to thinking about the place, functions, and interrelationships of objects: "OOO contends that nothing has special status, but that everything exists equally—plumbers, cotton, bonobos, DVD players, and sandstone, for example."[62] Triple O is developed around two contentions: the first is that "the assumption that only one subject—the human subject—is of [primary] interest or import"[63] is fallacious; the second, a consequence of the first, is "that objects do not relate merely through human use but through any use, including all relations between one object and any other."[64] As a result, the ambit of OOO is as ambitious as it is wide: "the objects of object-oriented thought mean to encompass anything whatsoever, from physical matter (a Slurpee frozen beverage) to properties (frozenness) to marketplaces (the convenience store) to symbols (the Slurpee brand name) to ideas (a best guess about where to find a 7-Eleven)."[65]

And yet the privileging of the agency of objects within object-oriented ontology has attracted criticism. For instance, in his chapter, Andrejevic argues that Ian Bogost's "embrace of a 'flat ontology' that espouses 'the abandonment of anthropocentric narrative coherence in favor of worldly detail'" can be read as "neatly parallel[ing] the anti-psychologistic, anti-narrative logic of drone sensing, and the analytic operations of the data mine." In contrast to such anthropocentric "abandonment," consideration for the human consequences of geocoded data is a key concern for Barreneche, Rossiter, and Andrejevic, as it is for a number of other contributors to this collection.

Political Economy, Politics, and Policy

As we have outlined, thinking and research on locative media have their genealogy in artistic and urban experimentation, as well as in the beginnings of military communications, mobiles, and telecommunications. Surprisingly, perhaps, the staple approaches in media and communication studies across political economy and policy have been slow to emerge in relation to locative media. There has been little critical work on the ownership and control of locative media. There is much policy interest, yet available research is limited and still clusters around privacy as the leading concern. So we are very pleased to include five chapters with a strong focus on these neglected areas (although a number of other contributions also do).

Rowan Wilken and Peter Bayliss's chapter on Foursquare provides a detailed, rigorous examination of the political economy of the marquee check-in application. Wilken and Bayliss give a careful overview and analysis of Foursquare and discuss its evolving business strategies and models. Especially helpful is the way that they situate Foursquare in the political economy of new media generally.

As Tarleton Gillespie points out, "much of the scholarship about the data collection and tracking practices of contemporary information providers has focused on the significant privacy concerns they provoke."[66] The coimplication of software and everyday spatial practices (as suggested by Kitchin and Dodge) further complicates these privacy concerns, raises policy issues, and forms the explicit focus of the chapters by Tama Leaver and Clare Lloyd, Timothy Dwyer, James Meese, and Gerard Goggin and César Albarrán-Torres.

In the first of four chapters on location privacy, Leaver and Lloyd examine a number of activist projects that set out to broaden consumer awareness of the information that is created from the use of locative media, so as to overcome what they term the "social media contradiction." Leaver and Lloyd define this as the situation where users of location-based social media applications tend to focus on the social elements and affordances of these systems, whereas the companies behind them focus on the media elements and commercial possibilities.

In the second of the privacy-focused chapters, Dwyer traces how the concept of privacy is changing rapidly as a result of remarkable growth in social media use and how our understanding of privacy is further complicated by the addition of location information. These developments lead Dwyer to contend that there is a pressing need for policy intervention, despite commercial pressures to exploit personal user data, as well as the need to push for the teaching of greater "tactical awareness" by end users.

Building on Dwyer's insights, in his Chapter 10, Meese looks at a specific locative media case study (the emergence of Google Glass) as viewed from within a particular jurisdictional context (Australia) and considers the likely legal and privacy implications that would follow the release of Glass in this country. Meese argues that Australia ought to introduce a statutory tort, alongside a broader program of legal education, in order to more fully respond to the privacy challenges posed by Google Glass and other locative media technologies.

In the fourth chapter that focuses on privacy, Goggin and Albarrán-Torres move beyond the Anglophone discussion with a case study of the Mexican Geolocalization Law. Examining the emergence of the Geolocalization Law in the currents of heightened societal concern with pervasive everyday crime, security arrangements, and the political dynamics of Mexico, Goggin and Albarrán-Torres argue for the need to understand the particular contexts and cultural specificity of locative media, as it is shaped in distinct places.

CONCLUSION

In conceiving this book, we set out to provide a set of ideas, resources, and frameworks for understanding locative media that seemed to be strangely absent at a time when the diffusion and interest in the technology were forging ahead and when conceptual innovation was occurring apace in cognate, interdisciplinary areas of Internet, mobile communications, and new media studies. We hope the assembled book helps to stimulate the kind of debate, inquiry, theory, and practical innovation and intervention we feel is needed in this important and still very much nascent area.

The gaps in the research field are quite clear. Most obviously, despite our own efforts in this volume, there is much to suggest that locative media are spreading and being adopted and adapted around the world. Yet the published research literatures, especially in English, do not reflect this trend. Rather, like many other areas of cultural, media, and communication studies, locative media research is modeled by imaginaries, assumptions, and standpoints from a restricted palette of countries, societies, sociodemographics, classes, and subcultures.

Encouragingly, we feel it is now the case that locative media research has moved from the reflex to associate this phenomenon with technical incubation, artistic experimention, noncommercial initiative, and urbanism and that, as this book reveals, researchers are now really engaging with the much wider range of instances, services, platforms, revenues, media ecologies, disconnections, and disarticulations that is the unruly emergence of locative media. We hope that research in political economy and policy, in particular, starts to "feel the width" of locative media because here, in particular, many more things are at play than just personal information and privacy (epochal those these might be).

NOTES

1. J. A. Simpson and E. S. C. Weiner, *The Oxford English Dictionary*, 2nd ed., Vol. VIII (Oxford: Clarendon Press, 1989), 1081–1082.
2. Simpson and Weiner, *The Oxford English Dictionary*, 1082.
3. Fred Lukermann, "The Concept of Location in Classical Geography," *Annals of the Association of American Geographers* 51 (1961): 194–210.
4. Lukermann, "The Concept of Location," 196.
5. Lukermann, "The Concept of Location," 201.
6. Lukermann, "The Concept of Location," 207.
7. Lukermann, "The Concept of Location," 197.
8. Ed Relph, *Place and Placelessness*, reprint and 3rd imprint (London: Pion, 1986), 3.
9. Adriana de Souza e Silva and Jordan Frith, *Mobile Interfaces in Public Spaces: Locational Privacy, Control, and Urban Sociability* (New York: Routledge, 2012), 9.
10. de Souza e Silva and Frith, *Mobile Interfaces*, 10.
11. Minna Tarkka, "Labours of Location: Acting in the Pervasive Media Space," in *The Wireless Spectrum: The Politics, Practices, and Poetics of Mobile Media*, eds. Barbara Crow, Michael Longford, and Kim Sawchuk (Toronto: University of Toronto Press, 2010), 131–145.
12. Tarkka, "Labours of Location," 134.
13. It is fitting, in this respect, that the *Oxford English Dictionary*'s third meaning of "locative"—"serving to locate or fix the position of something"—combines geography, navigation, and media technologies in its two examples of historical usage of this sense of the term. See Simpson and Weiner, *The Oxford English Dictionary*, 1082.
14. See Andrea Zeffiro, "A Location of One's Own: A Genealogy of Locative Media," *Convergence: The International Journal of Research into New Media Technologies* 18, no. 3 (2012): 249–266; Marc Tuters, "From Mannerist Situationism to Situated Media," *Convergence: The International Journal of Research into New Media Technologies* 18, no. 3 (2012): 267–282.
15. André Lemos, "Locative Media in Brazil," *Wi: Journal of Mobile Media* 10 (2009), http://wi.hexagram.ca/?p=60; Tarkka, "Labours of Location"; Christian Licoppe and Yoriko Inada, "Emergent Uses of a Multiplayer Location-Aware Mobile Game: The Interactional Consequences of Mediated Encounters," *Mobilities* 1, no. 1 (2006): 39–61.
16. Lucas Bambozzi, "Risky Approximations Between Site-Specific & Locative Arts," *Wi: Journal of Mobile Media* 10 (2009), http://wi.hexagram.ca/?p=56; Michael Salmond, "The Power of Momentary Communities: Locative Media and (In)Formal Protest," *Aether: The Journal of Media Geography* 5A (2010): 90–100, http://geogdata.csun.edu/~aether/pdf/volume_05a/salmond.pdf.

17. Tristan Thielman, "Locative Media and Mediated Localities: An Introduction to Media Geography," *Aether: The Journal of Media Geography* 5A (2010): 1–17, http://geogdata.csun.edu/~aether/pdf/volume_05a/introduction.pdf.
18. Alice Crawford and Gerard Goggin, "Geomobile Web: Locative Technologies and Mobile Media," *Australian Journal of Communication* 36, no. 1 (2009): 97–109; Adriana de Souza e Silva and Jordan Frith, "Locational Privacy in Public Spaces," *Communication, Culture & Critique* 3, no. 4 (2010): 503–525; Adriana de Souza e Silva and Daniel M. Sutko, "Theorizing Locative Media Through Philosophies of the Virtual," *Communication Theory* 21, no. 1 (2011): 23–42; Rowan Wilken, *Teletechnologies, Place, and Community* (New York: Routledge, 2011); Rowan Wilken and Gerard Goggin, eds., *Mobile Technology and Place* (New York: Routledge, 2012); K Willis, "Hidden Treasure: Sharing Local Information," *Aether: The Journal of Media Geography* 5A (2010): 50–62, http://geogdata.csun.edu/~aether/pdf/volume_05a/willis.pdf.
19. Sophia Drakopoulou, "A Moment of Experimentation," *Aether: The Journal of Media Geography* 5A (2010): 63–76, http://geogdata.csun.edu/~aether/pdf/volume_05a/drakopoulou.pdf; Alison Gazzard, "Location, Location, Location: Collecting Space and Place in Mobile Media," *Convergence: The International Journal of Research into New Media Technologies* 17, no. 4 (2011): 405–417; Francesco Lapenta, "Locative Media and the Digital Visualisation of Space, Place and Information," *Visual Studies* 26, no. 1 (2011): 1–3; Teri Rueb, "Shifting Subjects in Locative Media," in *Small Tech: The Culture of Digital Tools*, eds. Byron Hawk, David M. Rieder, and Ollie Oviedo (Minneapolis: University of Minnesota Press, 2008), 129–133.
20. Greg Elmer, "Locative Networking: Finding and Being Found," *Aether: The Journal of Media Geography* 5A (2010): 18–26, http://geogdata.csun.edu/~aether/pdf/volume_05a/elmer.pdf; see also Tarkka, "Labours of Location."
21. Eric Gordon and Adriana de Souza e Silva, *Net Locality: Why Location Matters in a Networked World* (Chichester, West Sussex: Wiley-Blackwell, 2011), 9.
22. Gordon and de Souza e Silva, *Net Locality*, 19.
23. Gordon and de Souza e Silva, *Net Locality*, 20.
24. Malcolm McCullough, "On the Urbanism of Locative Media," *Places* 18, no. 2 (2006), 26.
25. Gordon and de Souza e Silva, *Net Locality*, 20.
26. Gerard Goggin, *Global Mobile Media* (London: Routledge, 2011), 181.
27. Christoph Rosol, "From Radar to Reader: On the Origin of RFID," *Aether: The Journal of Media Geography* 5A (2010): 37–49, http://geogdata.csun.edu/~aether/pdf/volume_05a/rosol.pdf.
28. Melanie Swan, "Sensor Mania! The Internet of Things, Wearable Computing, Objective Metrics, and the Quantified Self 2.0," *Journal of Sensor and Actuator Networks* 1 (2012): 217–253.
29. Sensor technology and media and their social and policy implications were the subject of a May 2014 Defining the Sensor Society conference at the University of Queensland, http://cccs.uq.edu.au/sensor-society.
30. On ubiquitous computing, see Ulrik Ekman, ed., *Throughout: Art and Culture Emerging with Ubiquitous Computing* (Cambridge, MA: MIT Press, 2013); Paul Dourish and Genevieve Bell, *Divining a Digital Future: Mess and Mythology in Ubiquitous Computing* (Cambridge, MA: MIT Press, 2011); Anne Galloway, "Intimations of Everyday Life: Ubiquitous Computing and the City," *Cultural Studies* 18, nos. 2/3 (2004): 384–408; Mark Weiser, "Ubiquitous Computing," August 16, 1993, accessed March 5, 2014, www.ubiq.com/hypertext/weiser/UbiCompHotTopics.html.

31. See, for instance, Michael Batty, *The New Science of Cities* (Cambridge, MA: MIT Press, 2013); Brett Goldstein, ed., with Lauren Dyson, *Beyond Transparency: Open Data and the Future of Civic Innovation* (San Francisco: Code for America Press, 2013); Mark Deakin and Husam al Waer, eds., *From Intelligent to Smart Cities* (New York: Routledge, 2012); Adam Greenfield, *Against the Smart City*, part 1 of *The City Is Here for You to Use* (Kindle Digital Editions, 2013); Nicos Komninos, *The Age of Intelligent Cities: Smart Environments and Innovation-for-All Strategies* (New York: Routledge, 2014); Anthony Townsend, *Smart Cities: Big Data, Civic Hackers, and the Quest for a New Utopia* (New York: W. W. Norton, 2013).
32. Maurice Blanchot, "Everyday Speech," trans. Susan Hanson, *Yale French Studies* 73 (1987): 12–20.
33. Michael Foley, *Embracing the Ordinary: Lessons from the Champions of Everyday Life* (London: Simon & Schuster, 2012), 5.
34. Henri Lefebvre, *Critique of Everyday Life*, Vol. 1, trans. John Moore (London: Verso, 2000), 9–10.
35. Ben Highmore, *Ordinary Lives: Studies in the Everyday* (London: Routledge, 2011), 1.
36. Henri Lefebvre, "The Everyday and Everydayness," trans. Christine Levitch, Alice Kaplan, and Kristin Ross, *Yale French Studies* 73 (1987), 9.
37. Ben Highmore, *Everyday Life and Cultural Theory: An Introduction* (London: Routledge, 2002), 12. See, for example, Highmore, *Ordinary Lives*; Highmore, *Everyday Life and Cultural Theory*; Lefebvre, *Critique of Everyday Life*; Michel de Certeau, *The Practice of Everyday Life*, trans. Steven Rendall (Berkeley, CA: University of California Press, 1988); Michel de Certeau, Luce Giard, and Pierre Mayol, *The Practice of Everday Life*, Vol. 2: *Living and Cooking*, trans. Timothy J. Tomasik (Minneapolis: University of Minnesota Press, 1998); Michael Sheringham, *Everyday Life: Theories and Practices from Surrealism to the Present* (Oxford: Oxford University Press, 2006); Fran Martin, ed., *Interpreting Everyday Culture* (London: Hodder Arnold, 2003).
38. Highmore, *Ordinary Lives*, 115.
39. Rowan Wilken and Gerard Goggin, "Mobilizing Place: Conceptual Currents and Controversies," in *Mobile Technology and Place*, eds. Rowan Wilken and Gerard Goggin (New York: Routledge, 2012), 14. For a fine account of the importance of thinking through the issue of attention in relation to everyday life and our engagement with media and communications technologies, see Highmore, *Ordinary Lives*, 114–138. And for a somewhat different take on similar issues, one that draws from German media theory in developing the concept of "cultural techniques" as a way of making sense of human–technology interaction and "the links between human and non-human agencies," see Jussi Parikka, "Afterword: Cultural Techniques and Media Studies," *Theory, Culture & Society* 30, no. 6 (2013): 147–159.
40. Highmore, *Everyday Life and Cultural Theory*, 12.
41. Highmore, *Everyday Life and Cultural Theory*, 23.
42. Nigel Thrift, *Non-Representational Theory: Space, Politics, Affect* (London: Routledge, 2008), 2. As Thrift goes on to explain on the same page, he views nonrepresentational theory and his book as "the beginning of an outline of the art of producing a permanent supplement to the ordinary, a sacrament for the everyday, a hymn to the superfluous."
43. Thrift's account of how "sites" are understood from the perspective of nonrepresentational theories is interesting in that it builds on geographical work on relational understandings of space and place and incorporates Deleuzean understandings of "events." According to Thrift, nonrepresentational

theoretical approaches view the notion of the "site" "as an active and always incomplete incarnation of events, an actualization of times and spaces that uses the fluctuating conditions to assemble itself." Thrift, *Non-Representational Theory*, 12.

44. Hayden Lorimer, "Cultural Geography: The Busyness of Being 'More-Than-Representational,'" *Progress in Human Geography* 29, no. 1 (2005), 83.
45. Lorimer, "Cultural Geography," 84. Thus, as Ben Anderson and Paul Harrison point out, what "what pass[es] for representations [within nonrepresentational theory] are apprehended as performative presentations, not reflections of some *a priori* order waiting to be unveiled, decoded, or revealed." Ben Anderson and Paul Harrison, "The Promise of Non-Representational Theories," in *Taking-Place: Non-Representational Theories and Geography*, eds. Ben Anderson and Paul Harrison (Farnham, Surrey: Ashgate, 2010), 19.
46. Arjun Appadurai, *The Social Life of Things: Commodities in Cultural Perspective* (Cambridge: Cambridge University Press, 1986); Daniel Miller, *The Comfort of Things* (Cambridge: Polity Press, 2009); Nicky Gregson, *Living with Things: Ridding, Accommodation, Dwelling* (Wantage, Oxfordshire: Sean Kingston Publishing, 2007).
47. Harvey Molotch, *Where Stuff Comes From: How Toasters, Toilets, Cars, Computers and Many Other Things Come to Be as They Are* (New York: Routledge, 2003).
48. Sherry Turkle, ed., *Evocative Objects: Things We Think With* (Cambridge, MA: MIT Press, 2011); Richard Coyne, *The Tuning of Place: Sociable Spaces and Pervasive Digital Media* (Cambridge, MA: MIT Press, 2010).
49. John G. Palfrey and Urs Glasser, *Interop: The Promise and Perils of Highly Interconnected Systems* (New York: Basic Books, 2012).
50. See Jussi Parikka, "New Materialism as Media Theory: Medianatures and Dirty Matter," *Communication and Critical/Cultural Studies* 9, no. 1 (2012): 95–100; Rick Dolphijn and Iris van der Tuin, eds., *New Materialism: Interviews and Cartographies* (Ann Arbor, MI: Open Humanities Press, 2012); Diana Coole and Samantha Frost, eds., *New Materialisms: Ontology, Agency, and Politics* (Durham, NC: Duke University Press, 2010).
51. Diana Coole and Samantha Frost, "Introducing the New Materialisms," in *New Materialisms: Ontology, Agency, and Politics*, eds. Diana Coole and Samantha Frost (Durham, NC: Duke University Press, 2010), 2.
52. Coole and Frost, "Introducing the New Materialisms," 5.
53. Jeremy Packer and Stephen B. Crofts Wiley, "Introduction: The Materiality of Communication," in *Communication Matters: Materialist Approaches to Media, Mobility and Networks*, eds. Jeremy Packer and Stephen B. Crofts Wiley (London: Routledge, 2012), 12. See also Tarleton Gillespie, Pablo J. Boczkowski, and Kirsten A. Foot, eds., *Media Technologies: Essay on Communication, Materiality, and Society* (Cambridge, MA: MIT Press, 2014).
54. Adriana de Souza e Silva and Jordan Frith, "Location-Aware Technologies: Control and Privacy in Hybrid Space," in *Communication Matters: Materialist Approaches to Media, Mobility and Networks*, eds. Jeremy Packer and Stephen B. Crofts Wiley (London: Routledge, 2012), 273.
55. Richard Coyne, *The Tuning of Place: Sociable Spaces and Pervasive Digital Media* (Cambridge, MA: MIT Press, 2010).
56. Rob Kitchin and Martin Dodge, *Code/Space: Software and Everyday Life* (Cambridge, MA: MIT Press, 2011), 16. The fuller implications of this mutual constitution of software and everyday life have been explored elsewhere by numerous scholars. For key works, see Adrian Mackenzie, *Cutting Code: Software and Sociality* (New York: Peter Lang, 2006); David M. Berry, *The*

Philosophy of Software: Code and Mediation in the Digital Age (Houndmills, Basingstoke, Hampshire: Palgrave Macmillan, 2011); Tarleton Gillespie, "The Relevance of Algorithms," in *Media Technologies: Essays on Communication, Materiality, and Society*, eds. Tarleton Gillespie, Pablo Boczkowski, and Kirsten A. Foot (Cambridge, MA: MIT Press, 2014); Noah Wardrip-Fruin, *Expressive Processing: Digital Fictions, Computer Games, and Software Studies* (Cambridge, MA: MIT Press, 2009).

57. Martin Dodge, Rob Kitchin, and Matthew Zook, "How Does Software Make Space? Exploring Some Geographical Dimensions of Pervasive Computing and Software Studies," *Environment and Planning A* 41, no. 6 (2009), 1284.
58. Dodge, Kitchin, and Zook, "How Does Software Make Space?" 1284.
59. Carlos Barreneche, "The Order of Places: Code, Ontology and Visibility in Locative Media," *Computational Culture: A Journal of Software Studies* 2 (2012), http://computationalculture.net/article/order_of_places; Carlos Barreneche, "Governing the Geocoded World: Environmentality and the Politics of Location Platforms," *Convergence: The International Journal of Research into New Media Technologies* 18, no. 3 (2012): 331–351.
60. Parikka, "New Materialism as Media Theory," 96.
61. Bruno Latour, *Reassembling the Social: An Introduction to Actor-Network-Theory* (Oxford: Oxford University Press, 2005).
62. Ian Bogost, *Alien Phenomenology, or What It's Like to Be a Thing* (Minneapolis: University of Minnesota Press, 2012), 6.
63. Bogost, *Alien Phenomenology*, 23.
64. Bogost, *Alien Phenomenology*, 6.
65. Bogost, *Alien Phenomenology*, 23–24.
66. Gillespie, "The Relevance of Algorithms," 173.

Part 1

Practices, Publics, Spaces

2 Intimate Cartographies of the Visual

Camera Phones, Locative Media, and Intimacy in Kakao

Larissa Hjorth

With the rise of high-quality camera phones, accompanied by the growth of in-phone editing applications and distribution services via social and locative media, we are witnessing new types of co-present visuality. In the first series of studies of camera phones by the likes of Mizuko Ito and Daisuke Okabe,[1] they noted the pivotal role played by the three "S's"—sharing, storing, and saving—in informing the context of what was predominantly "banal" and intimate everyday content. For Ilpo Koskinen, camera phone images were branded by their participation in a new type of banality.[2] Although this banality can be seen as extending the conventions and genres of earlier photographic tropes (that is, associated with analogue photography),[3] they also significantly depart.

As camera phones become more commonplace in the explosion of smartphones—along with new contexts for image distribution like microblogging and location-based services (LBS)—emergent types of visual overlays become apparent. Through the deployment of geotagging, these visualities are geotemporally linked to locations. In these overlays, the social is linked to the electronic and the geographic to the temporal—giving way to new forms of media and mediated "cartographies" that impact an experience of a locality. With the move beyond thinking about the representation of place through two-dimensional maps, we can see such cartographies consisting of overlays between the temporal, spatial, social, and electronic. Images are given ambient, networked contexts in which the geographic is overlaid with the social and emotional.

While, globally, camera phone genres like self-portraiture have blossomed, we are also witnessing the flourishing of vernacular visualities that reflect a localized notion of place, sociality, and identity-making practices.[4] Smartphone apps like Instagram have made taking and sharing photographs easier and more compelling, creating new overlays between image, place, and intimacy. With its retro analogue-looking filters, Instagram epitomizes the dictum that new media are haunted by the old. Although Instagram's geotemporal tagging links the geographic with the temporal, it "suppresses temporal, vertical structures in favor of spatial connectivities."[5] In short, it emphasizes a particular dimension of place—for example, "location" as spatially determined—while evoking esthetics of yesteryear and nostalgia.[6]

Given this reframing of the camera phone and its relationships with image, movement (across spatial and temporal realms), and intimacy through LBS, this chapter explores a case study of users famous for their camera phone usage: Korean women.[7] South Korea offers a fascinating example not only because it has played such an important role in camera phone practices historically and boasts some of the highest smartphone adoption rates in the world,[8] but also because it is also home to key global mobile media companies like Samsung and LG. Korean female users have a long history as early adopters of camera phone taking and sharing, especially through *sel-ca* (self-portraiture).

In particular, this chapter draws on case studies of Korean female respondents and their use of the first purpose-built social mobile media platform, Kakao (Talk, Story, Place), undertaken as part of a three-year ethnographic study into online media in the Asia-Pacific region from 2009 to 2012. Over the three years, I conducted fieldwork for a few months every year in each location. Fieldwork consisted of participant observation, focus groups, in-depth interviews, and scenarios of use diaries. By the end of 2012, a shift in media usage had emerged whereby gendered, generational, and cultural differences became apparent.[9] In this chapter, I discuss gendered differences by drawing on a case study of 30 female users between the ages of 20 and 50 years conducted in 2013. This case study is meant not to be demonstrative of all South Korean practices but rather to provide insight into some of the experiences and nuances informing locative mobile media practice.

Unlike media such as Facebook, which was built for PC use and has had to adapt to the smartphone becoming the key (if not only) portal for online applications, Kakao was built as a series of free messaging and talking platforms that also featured a social media space (with cyber room) and a suite of hundreds of mobile games including mobile multiplayer online games like *I Love Coffee*. Exploring practices connected with the emergence of Kakao not only provides us insight into how purpose-built mobile social media has particular affordances but also gives us understanding of how a locality shapes those affordances. And yet, unlike many of the mobile media apps and platforms where location-based services are set almost by default, Kakao's various apps have been less locative media–centric. Far from uniform in characteristics and texture, Kakao highlights how we might begin to understand the complex ways in which locative, social, and mobile media convergence is far from even. By focusing on women, a demographic most explicitly impacted by locative media in terms of security and risks, this chapter seeks to understand locative media as not a default for contemporary mobile media but rather as a nuanced space that involves resistances as well as uptake.[10]

This chapter explores the way in which camera phone picture taking and sharing can be viewed as part of a broader process of "emplacement" within everyday movements. By "emplacement," I refer to the ways in which notions of place are framed through engagements across localities. With the decline in *sel-ca* in Korea, new types of image making and representations of copresent places can be seen. This chapter first contextualizes the changes in camera phone practices with the rise of locative, social, and mobile media convergence. I turn then to the case study, detailing the Kakao social mobile media platform, and then to a discussion of the use (and nonuse) of camera phones and LBS by respondents interviewed in 2013.

SNAPSHOTS IN MOTION: SHIFTS IN CAMERA PHONE PRACTICES

As Daniel Palmer notes in his study on iPhone photography, "cameras have colonized the mobile phone over the past decade."[11] Nokia has reportedly put more cameras into people's hands than in the whole previous history of photography.[12]While camera phone genres such as self-portraiture have blossomed on a global scale, vernacular visualities that reflect a localized notion of place, sociality, and identity-making practices[13] are also flourishing. Smartphone apps like Snapchat and Instagram have made taking and sharing photographs easier and more compelling.[14] Palmer notes that Snapchat (an app and photo-sharing service whose primary innovation is the self-destructing image, aka the anti-Kodak moment) focuses on a "pornographication of experience" with the "commodification of surprise."[15] As usual, the cultural problem is not necessarily "the exposure of the previously private" but the erosion of the boundary between public and private spaces.[16]

With location-based social networking services like KakaoPlace and location-based games like Foursquare and Jiepang, we see a further overlaying of location with the social and personal, whereby the electronic is superimposed onto the geographic in new ways. Specifically, by sharing an image and comment about a place through LBS, users can create different ways to experience and record journeys and, in turn, affect how a location is recorded, experienced, and thus remembered. In this ambient, intimate, and mobile visuality, movement across geographic, temporal, social, and emotional spaces comes to the forefront in what can be characterized as "emplaced cartographies."[17]

This shift is especially the case with the overlaying of ambient images within moving narratives of place across various localities as afforded by LBSs. An example might be someone uploading a geotagged camera phone image onto Kakao, whereupon KakaoPlace adds various geographic, temporal, and spatial details about the location that then informs the user how that location is recorded and shared. This practice, in turn, impacts on their experience of location as something that is mediated through networked media. Although a sense of place and intimacy has always been mediated by

HUMBER LIBRARIES

language, memories, and gestures, it is the way in which it is being mediated that is transforming how we think about and experience a location and its relationship to intimacy.

The rapid uptake of smartphones has enabled new forms of distribution and provided an overabundance of apps, filters, and lenses to help users create "unique" and artistic camera phone images. Although the iPhone has been quick to capitalize on this phenomenon through applications such as Instagram and Snapchat, other operating systems like Android have also had their share of this expanding market. So, too, social media, such as microblogs and LBSs, have acknowledged the growing power of camera phone photography, not only by affording easy uploading and sharing of the vernacular[18] but also by providing filters and lenses in order to further enhance the "professional" and "artistic" dimensions of the photographic experience.[19]

Consider, for example, how social media applications for smartphones no longer ask you to go through multiple steps to attach images to a post. Formerly, if you wanted to post an image, you would take the image using the phone's camera application that would store the image in the camera's library, and you would then access the shot by attaching the image to a post. Sometimes you would even need to upload the image to an online image repository so that it could be linked to the SNS. Now, many social media apps provide a photo button integrated into the app that allows you to take a picture and post it immediately, and social media companies provide their own image-hosting servers that operate almost invisibly to the user. One way to understand the impact of LBS on camera phone practices is through a shift from networked to emplaced visuality.

When visuality becomes part of a networked culture, its meanings, contexts, and content change. Camera phones, as an extension of the networked nature of mobile media, are clearly defined by this dynamic. Although initial studies into camera phone visuality discussed it as part of "networked media,"[20] this second generation of visuality—one that is characterized by locative media—is about emphasizing the role of movement as central to our experiences of a location.[21] These exercises are emotional and electronic, geographic, and social—highlighting the complexity of ever evolving notions of place. In each location, camera phone images are overlaid onto specific places in a way that reflects existing social and cultural intimate relations, as well as being demonstrative of new types of what Sarah Pink calls "emplaced" visuality, in which locative media emplace images within the entanglement of movement.[22]

First-generation "networked" visuality, when combined with LBSs in the obvious case of the smartphone, becomes "emplaced" visuality, that is, visuality mapped by a moving, geospatial sociality. Incorporating movement in the theorization of visuality is important given the ways in which camera phone practices give way to an accelerated taking, editing, and sharing of a "moment" that is then contextualized through its place in the moving geographic and social maps of LBSs and social media. Whereas first-generation camera phone sharing was defined by the network,[23] the second generation—characterized by the geotemporal features of geotagging—becomes focused *on emplacement through movement.*[24] In other words, the LBS camera phone culture is about reinforcing the process of the node rather than the product of the network. Images increasingly became about creating a sense of movement though an ambience of place. These images are "multisensorial" in that they evoke more than the visual; they overlay information (such as location) with emotion.[25]

For Pink, the combination of locative media with the photographic image requires a new paradigm that engages with the multisensoriality of images. It might at first seem odd to talk about images as being multisensorial because surely images are visual and so draw upon only one sense: vision. Pink draws on Tim Ingold's critique[26] of the anthropology of the senses and of network theory, arguing that by exploring the visual in terms of multisensoriality, one can reprioritize the importance of movement in understanding a sense of locality. For Pink, locative media provide new ways in which to frame images with the "continuities of everyday movement, perceiving, and meaning making."[27]

By contrasting "photographs as mapped points in a network" with "photographs being outcomes of and inspirations within continuous lines that interweave their way through an environment—that is, in movement and as part of a configuration of place,"[28] Pink argues that we must start to conceive of images as produced and consumed in movement. This shift can be viewed as the movement from a camera phone visuality that is networked to camera phone images that are "emplaced."[29] An image that is socially networked, tagged, and GPS located is "emplaced" in a number of ways—through social connectivities, through copresence, through geotemporal tagging.

Finally, the social distribution of the images creates a social public for those images, thus overlaying another context, and the image tags entered by the public overlay yet another context. As Mikki Villi notes in "Visual Mobile Communication on the Internet: Publishing and Messaging Camera Phone Photographs Patterns," "much of the traffic in photographs now circulates through digital networks and is facilitated by new platforms."[30] And, as Lapenta notes, "what is really changing has little to do with the increasing numbers of images taken every day and more to do with the increasingly differentiated forms of photographic image production, aggregation and distribution."[31]

CASE STUDY: SOCIAL MOBILE MEDIA AND KAKAO

A few years ago, Soohyun would have been preoccupied with South Korea's first social network site, Cyworld minihompy. She and her friends would often meet in their online mini rooms and exchange virtual gifts symbolic of their friendship. Now Soohyun uses Kakao, a mobile social media platform. KakaoTalk allows her to text and talk for free, includes an online chat function similar to a minihompy, and features a whole suite of mobile games, from the multiplayer cafe-simulation game *I Love Coffee* to the retro-cute match-three puzzle app Anipang. Significantly, unlike Facebook and other social media services that provide reconfigured mobile apps across devices and operating systems, KakaoTalk is designed specifically for mobile media and, more specifically, for smartphones, enabling a targeted tailoring of the platform to the particular social, ludic, and sensory affordances of mobile screens. For people like Soohyun, the choice of Kakao was obvious. It afforded her integrated modalities of presence—polysynchronous, distributed, and ambient—with friends and family across both communicative and playful mobile practices: texting, talking, online chat, and gaming. It has also afforded her more personalization regarding when to use and not use LBS.

Created by Beom-Soo Kim, Kakao was released in march 2010. By December 2012, it had 70 million registered users with over half of them operating the app daily.[32] More than 90 percent of Korean smartphone users predominantly use KakaoTalk, and 3.4 billion messages are sent everyday. According to one marketing consulting firm, Kakao has become part of the South Korean vernacular:

> Need to find information? "Google it." Need an overnight delivery? "FedEx it." In Korea, "Send me a message" is rarely heard. Instead, people say "Katalk me."[33]

Initially popular in Korea, Indonesia, and Vietnam (in the wake of the Gangnam Style success), KakaoTalk is now seeking to grow its user base through customized iPhone and Android apps and to position itself as "a Mobile Social Platform Pioneer."[34] In addition to supporting free one-to-one, group, and conference calls, KakaoTalk also enables the sharing of photos, videos, and contact information (with no banner ads). Exploiting the reality of hectic Korean life that boasts some of the longest work hours under highly competitive conditions, KakaoTalk allows friends and family to catch up online across various platforms and modalities. Through KakaoTalk's instant messaging service, Kakao has been quick to gain millions of registered users who have then migrated to the various other services offered by the social mobile media platform. By meshing social, locative, and playful mobile media in new ways, KakaoTalk provides some insight into the cultural and gender specificities of mobile gaming; indeed, Kakao has been particularly successful at engaging the attention of millions of young women through their social media game, *I Love Coffee*.

As the first mobile platform specific to the provision of a rich and multimodal social media experience, Kakao has combined a variety of media experiences and practices, all focused on strengthening social relationships. Since 2012, KakaoTalk has enhanced and further integrated its social, locative, and mobile media elements with the release of KakaoStory (a "place" to upload photos and share "daily life stories," with in-app filters, emoticons, and "photo wall" capability); Choco (Kakao currency or "cyber money" for purchasing emoticons and other content); KakaoGame (a collection of mobile games that enable users to share their scores with friends); and KakaoStyle (information on the latest fashion trends). Within the first eight days, KakaoStory attracted 10 million users. The Kakao mobile app now has more than 30 million users, outnumbering offerings from Facebook (9.49 million) and Twitter (6.42 million) in South Korea.[35]

Part of Kakao's success has been its ability to integrate a variety of services from communication and socializing to gaming and content sharing. Drawing on Korea's love of online games, it has effectively shifted the focus on massively multiple online games (MMOs) played in PC *bangs* (PC rooms) to mobile games by securing the casual and yet highly committed female players. Moreover, by deploying consistent user interface design and simple, familiar graphics among social, mobile, and locative media, Kakao has been able to develop its own version of Korea's longest running social media cyber room.

Adapting Cyworld's minihompy, which was a multimodal interface—game hub, cyber room, online archive, camera phone journal, music room, and social network site[36]—Kakao has been quick to capitalize on the smartphone market. Whereas Cyworld's mobile apps had offered a glimpse of what was fully available online (via a personal computer), Kakao has focused on providing a suite of services that take full advantage of the mobile phone as a web browser, game device, social networking tool, and location-based media form. In the next section, I discuss the use of Kakao, camera phone sharing, and LBS in movements across different locations.

AMBIENT LOCALITIES: KAKAO IMAGE SHARING AND PLAY

The rise of camera phone geotagging has provided new ways for users to play with sociality, copresence, and location. Here we are reminded of Brian Sutton-Smith's work in the area of play as a vehicle for understanding the relative and sociocultural dimensions of localities.[37] In the face of growing cultural ludification, whereby numerous elements of cultural practice have been commodified through game play, it is increasingly important to look at play as distinct from the game play conflation. Camera phone sharing is one example of playing with ambience and location. Often camera phone sharing occupies the unofficial role in and around LBS practices. In the case of the various services offered by Kakao, female respondent Sol has found that her social media use—as well as, more generally, her play—has increased in both dimension and frequency. She notes that KakaoStory has become an important portal for socializing and sharing camera phone pictures and stories while on the move:

> I use more social media than before. I check my email and KakaoStory feeds when I'm having lunch at a café or traveling in the subway. When I'm alone, my mobile phone is a friend. I can chat with my friend through KakaoTalk even though we are not in the same place. So I use a lot of social media through my smart phone. And I use more camera phone apps too. I use KakaoStory a lot these days. It is easy to take and upload photos. I also like to see my friends' KakaoStory photos. Every morning when I wake up, I check my KakaoStory feeds and reply to my friends' comments about my photos. Most of my friends don't use Cyworld anymore so I moved to KakaoStory to communicate with my friends. If my friends don't use Cyworld or any kinds of social media, there is no point to use that. I use social media to see my friends.

The integration of photos as a fundamental—and often primary—mode of messaging and networking is one of the most significant trajectories of camera smartphone functionality and part of the broader convergence of media and communication in contemporary culture (i.e., communicating *with* our own "small media" content). Theorists of visual culture have argued that camera phones—due in part to their corporeal intimacy (always in hand or close to the body) and ever presence—have fundamentally impacted upon our visual literacy and reconfigured the "photographer–technology image relationship,"[38] changing the way we use and experience photographic images and our relation to that which is "photographable."

Daisuke Okabe and Mizuko Ito argue the mobile phone has three central properties: it is pedestrian (on-the-street, pervading all settings and locations), portable (on-the-body, both in the home setting and outside), and personal (literally both a self-portal and private archive).[39] Appropriating these properties, the phone camera's ubiquitous visual access effectively heightens our visual awareness of the everyday, converting every situation into a potential photo or mini video narrative opportunity.[40] This kind of user-generated photographic content and practice—exemplified by Sol's use of KakaoStory—is deeply embedded in the way we use social media and networking services, such that photography becomes interwoven with, and modified by, other patterns of mobile communication and social "play."[41]

I asked Sol, as an active taker and sharer of camera phone pictures through KakaoStory, to provide some examples of the types of pictures she took and the stories behind them. In the first picture (Figure 2.1), she describes her visit to a nail salon where she got the letters of her name painted on her nails, how she enjoyed the moment, and wanted to share it immediately—in the moment—with friends. She received many responses and said some of

Figure 2.1 Sol's KakaoStory, image 1.

her friends also decided to do the same and subsequently uploaded their photos; together, they participated in a telepresent and asynchronous form of collective or shared action.

In the second picture (Figure 2.2), Sol shares a moment of absentmindedness with friends, showing them how she had accidently put her television remote into her bag that morning. As she states:

> I don't know why I grabbed and brought it . . . Maybe I need to go to see a doctor to check my brain. People made comments such as, "Yes, go and check . . . Crazy!" "What's wrong with you?" "As usual . . . ," etc. I took this photo because it was weird. I was so tired and hadn't slept well. Actually I couldn't sleep at all, I worked the whole night . . . After I came out of my house and when I locked the door, I realized that I grabbed a TV remote control!!! This is weird and I was sure that my friends would laugh at me so I uploaded this photo for fun.

Figure 2.2 Sol's KakaoStory, image 2.

For Sol and her friends, sharing the private, mundane, and everyday aspects of their lives via KakaoStory represents a merging of online and offline modes of being-in-the-world, where an immediate and embodied "happening" stretches into the network via the communicative act of uploading mobile media content onto a shared mobile web interface. Okabe and Ito suggest that this kind of sharing facilitates a sense of connection defined by what they call ambient visual copresence.[42] Although Sol and her friends are not physically together, by uploading, messaging, and sharing each other's day-to-day experiences, they can visually tap into and feel part of their friendship group in a diffuse and polychromatic manner.

As noted in previous work about Cyworld's minihompy, the official virtual currency and unofficial forms of gift giving played an essential role in the ongoing success of the social media space (for over a decade).[43] As these gift-giving practices circumnavigate mobile, social, and locative worlds with increasing complexity, the relationship between modes of presence and place and the various networked pathways of communication take on new significance as emergent forms of "mobile intimacy."[44]

For Jina, the motivations for sharing camera phone images via social mobile media are about sharing experiences and feelings almost instantaneously with friends elsewhere across both online and offline spaces, a way of negotiating and traversing one's being-in-the-world with what has been termed the "absent presence" central to the rise of mobile communication.[45] As Jina notes:

> I usually upload photos taken with my friends. I want to share the feeling and the moments with my friends so I took and uploaded it. Or if I want to share some useful information with my contacts, I put images and comments on my wall or KakaoStory . . . I spend a lot of time using social media with my smartphone—from Instagram, KakaoStory, Path, and Foursquare. It's part of my daily life.

For Kyung, smartphones have also accelerated the amount and frequency of her social media use. She has witnessed a rapid rise in camera phone sharing via KakaoStory as a way in which to be copresent on the move. For Kyung, images can be more evocative and playful, providing more space for responses. Like many of the other respondents, Kyung no longer used Cyworld because its interface was not nuanced for smartphones. As she states:

> I don't like the interface of Cyworld anymore. KakaoStory is very easy to use with a smartphone. And I prefer KakaoStory because I can share my photos and stories with my "real" friends, not "unspecified individuals." It is fun and makes me keep in more constant contact with close friends.

Kyung highlights the highly customized nature of KakaoStory for just sharing with real friends. For Kyung, camera phones are an important part of narrating life with, and for, friends. For Kyung, KakaoStory allowed her to create and share her own "small media" content through online visual journals. As she notes:

> When I travelled to India last year, I took photos and uploaded them on my KakaoStory account. I uploaded photos during my journey because I just wanted to share my feelings, places where I've been and foods that I've eaten during my Indian trip with my friends in Korea. So I did "taking, uploading and sharing" photos. I upload photos often because I want to share the moment of what I've seen with my friends. I think "uploading photos" means "sharing photos." If I want to keep something as a secret, I won't upload. I use geotagging on Facebook (there's no geotagging on KakaoStory) when I go somewhere I've never been. Just for fun . . . And sometimes I don't want to put my current location on my Facebook, but it does automatically. Then I feel my private information is exposed too much.

Kyung's view of uploading as synonymous with sharing is an important one. Here we see that geotagging for Kyung is useful only when traveling to clearly align particular experiences with locations. Otherwise she prefers not to geotag so that the images taken on the move within everyday life are not limited to just the geographic. The lack of a geotagging default on KakaoStory (tagging happens through the specific app KakaoPlace) is important for Kyung in motivating her to continue to take camera phone pictures within the movements of everyday life. As Kyung states:

> Geotagging is good when you are travelling to help organize events and memories and link them specifically to a place. But in daily life they give too much information that I don't want everyone to know. I might be happy for my friends to know where I am when I take a picture but if you upload you share with friends and strangers. It's the strangers I don't want to know [geographic] information about my activities.

For Yun, as a mother of two young children, social media played a key role in documenting their lives and sharing it with friends and family. As she notes:

> I mostly take my two children's photos. I like to share my boys with my friends and love to read my friends' comments about my boys. But actually, I take, upload photos, and write short comments regularly, for "recording." It is more like my "baby diary." But I don't like using geotagging. I think it's enough that I am creating an image archive of my children online without giving everybody information about them. I also like when friends say, "oh, that looks like an interesting place" and then I can reply and tell them where it is and we can have a conversation. Giving too much information doesn't allow for conversation.

The use of camera phone practices to create a data archive for personal memories, moments, and stories is becoming increasingly prevalent. Here Yun highlights the ways in which geotagging, by attaching a moment to a specific location, limits the potential for conversation. Yun's social media diet primarily included Cyworld–Egloo (a Korean blog, similar to blogspot) and KakaoStory. For Yun:

> I move to new social media depending on the number of users (popularity of social media). Or, if I don't want to show my private life anymore, I just move to another new social media and start over again. If I block someone or delete someone, they will be angry about this, so I just move to another social media rather than deleting people whom I don't want to show my privacy any more.

Here Yun reminds us of the multiple dimensions of privacy at play.[46] Often, individual notions are informed by cultural and regulatory understandings of privacy. Although historically the importance of collectivism has influenced practices such as privacy, with the rise of cases such as the Samsung illegal tracking of workers[47] and government/corporate cronyism,[48] Korean uses have become increasingly mindful of the negative dimensions of the Information Age.[49] Echoing danah boyd's discussion of generational differences in understanding privacy,[50] in Korea we see even more gradations of understanding across generational and class differences. The uptake of KakaoStory, one of the few social mobile media that does not have locative media usage as a default setting, highlights some of these nuanced and localized notions of privacy and how they inform media practices more generally.

CONCLUSION: MAPPABLE INTIMACIES

Although second-generation LBS games like Foursquare and Jiepang are in their infancy, they represent an area of growing diversity and complexity within mobile media and communication. However, outside the magic circle of these games, unofficial forms of play can be found through camera phone taking and sharing. Increasingly, ambient play comes to the forefront. This is particularly the case in the growth and sustained interest in camera phone practices, with the shift to more complex geotemporal overlays with geotagging. Camera phone practices have historically been gendered activities, with early studies highlighting the importance they played in women's representation and also in notions of empowerment.[51]

With geotagging added into the mix, how we visualize the intimate cartographies of camera phone practices is changing. As I have suggested in this chapter, the notion of emplacement provides a more accurate sense of the changing relationship between movement, place, and location than the notion of networked does, and thus it can be useful in understanding new forms of LBS camera phone practices.

The mainstreaming of LBSs through smartphone visuality—vis-à-vis camera phone apps like Instagram—is demonstrating the diverse ways in which sociality and intimacy are overlaid on the geographic and spatial as part of everyday movement. In this chapter, I have explored the uptake and resistance to LBS around the ambient play of camera phones through a case study of Kakao in South Korea. I have argued that if we are to truly understand the ways in which locations shape media, and vice versa, we need to contextualize use with nonuse. Kakao, as the first purpose-built mobile social media platform, provides an interesting context in which to examine the uptake of LBS within particular media ecologies. Future studies in locative media will need to reflect on these issues of use and nonuse within and across platforms and apps if we are to fully understand the impact and resistance to locative media within the messy space of the everyday.

NOTES

1. Mizuko Ito and Daisuke Okabe, "Camera Phones Changing the Definition of Picture-Worthy," *Japan Media Review* (August 29, 2003), accessed January 10, 2013, www.ojr.org/japan/wireless/1062208524.php; Mizuko Ito, "Intimate Visual Co-Presence" (paper presented at the Ubiquitous Computing Conference, Tokyo, Japan, September 11–14, 2005), accessed January 10, 2013, www.itofisher.com/mito/publications/intimate_visual.html; Mizuko Ito and Daisuke Okabe, "Everyday Contexts of Camera Phone Use: Steps Towards Technosocial Ethnographic Frameworks," in *Mobile Communication in Everyday Life: An Ethnographic View*, eds. Joachim Höflich and Maren Hartmann (Berlin: Frank & Timme, 2006), 79–102.
2. Ilpo Koskinen, "Managing Banality in Mobile Multimedia," in *The Social Construction and Usage of Communication Technologies: European and Asian Experiences*, ed. Raul Pertierra (Singapore: Singapore University Press, 2007), 48–60.
3. Larissa Hjorth, "Snapshots of Almost Contact," *Continuum*, 21, no. 2 (2007): 227–238.
4. Dong-Hoo Lee, "Women's Creation of Camera Phone Culture," *Fibreculture Journal* 6, (2005), accessed July 11, 2012, www.fibreculture.org/journal/issue6/issue6_donghoo_print.html; Larissa Hjorth, "Re-Imaging Urban Space: Mobility, Connectivity, and a Sense of Place," *Mobile Technologies*, eds. Gerard Goggin and Larissa Hjorth (London: Routledge, 2009), 235–251; Hjorth, "Snapshots of Almost Contact."
5. Nadav Hochman and Lev Manovich, "Zooming into an Instagram City: Reading the Local Through Social Media," *First Monday* 18, no. 7 (2013), accessed July 5, 2013, http://firstmonday.org/ojs/index.php/fm/article/view/4711.
6. See Didem Ökzul's work for a detailed discussion of locative media, place, and the politics of nostalgia.
7. Lee, "Women's Creation of Camera Phone Culture."
8. Nielsen KoreanClick, *May Report* (2012), accessed July 11, 2012, www.koreanclick.com/.
9. Larissa Hjorth and Michael Arnold, *Online@Asia-Pacific* (New York: Routledge, 2013).
10. Katie Cincotta and Kate Ashford, "The New Privacy Predators," *Women's Health* (2011), accessed January 10, 2013, www.purehacking.com/sites/default/files/uploads/2011_11_00_Australian_Womens_Health_November.pdf.
11. Daniel Palmer, "Mobile Media Photography," *The Routledge Companion to Mobile Media*, eds. Gerard Goggin and Larissa Hjorth (New York: Routledge, 2014).
12. This claim appears on Nokia's website in "Camera Phones Backgrounder," accessed June 20, 2011, www.nokia.com/NOKIA_COM_1/Microsites/Entry_Event/Materials/Camera_phones_backgrounder.pdf.
13. Hjorth, "Snapshots of Almost Contact"; Lee, "Women's Creation of Camera Phone Culture"; Hjorth, "Re-Imaging Urban Space: Mobility, Connectivity, and a Sense of Place."
14. Chris Chesher, "Between Image and Information: The iPhone Camera in the History of Photography," *Studying Mobile Media: Cultural Technologies, Mobile Communication, and the iPhone*, eds. Larissa Hjorth, Jean Burgess, and Ingrid Richardson (London/New York: Routledge, 2012), 98–117; Daniel Palmer, "iPhone Photography: Mediating Visions of Social Space," *Studying Mobile Media: Cultural Technologies, Mobile Communication, and the*

iPhone, eds. Larissa Hjorth, Jean Burgess, and Ingrid Richardson (London/New York: Routledge, 2012), 85–97.
15. Palmer, "Mobile Media Photography."
16. Scott McQuire, *The Media City: Media, Architecture and Urban Space* (London: Sage, 2008), 204.
17. Sarah Pink and Larissa Hjorth, "Emplaced Cartographies: Reconceptualising Camera Phone Practices in an Age of Locative Media," *Media International Australia* 145 (2012): 145–156.
18. Jean E. Burgess, "Vernacular Creativity and New Media" (PhD diss., Queensland University of Technology, 2007), accessed July 5, 2013, http://eprints.qut.edu.au/16378/.
19. Søren Mørk Petersen, "Common Banality: The Affective Character of Photo Sharing, Everyday Life and Produsage Cultures" (PhD diss., IT University of Copenhagen, 2009).
20. Daniel Rubinstein and Katrina Sluis, "A Life More Photographic: Mapping the Networked Image," *Photographies* 1, no. 1 (2008): 9–28; Mikko Villi, "Visual Mobile Communication on the Internet: Publishing and Messaging Camera Phone Photographs Patterns," *Seamlessly Mobile*, eds. Kathleen Cumiskey and Larissa Hjorth (New York: Routledge: 2013); Ito and Okabe, "Camera Phones Changing the Definition of Picture-Worthy"; "Intimate Visual Co-Presence"; "Everyday Contexts of Camera Phone Use."
21. Pink and Hjorth, "Emplaced Cartographies."
22. Sarah Pink, "Sensory Digital Photography: Re-Thinking 'Moving' and the Image," *Visual Studies* 26 vol. 1 (2011): 4–13.
23. Ito and Okabe, "Everyday Contexts of Camera Phone Use"; Villi, "Visual Mobile Communication on the Internet"; Burgess, *Vernacular Creativity*.
24. Pink, "Sensory Digital Photography."
25. Pink, "Sensory Digital Photography."
26. Tim Ingold, "Anthropology Is Not Ethnography," *Proceedings of the British Academy*, 154 (2008): 69–92.
27. Pink, "Sensory Digital Photography."
28. Pink, "Sensory Digital Photography."
29. Pink, "Sensory Digital Photography."
30. Villi, "Visual Mobile Communication on the Internet."
31. Francesco Lapenta, "Geomedia: On Location-Based Media, the Changing Status of Collective Image Production and the Emergence of Social Navigation Systems," *Visual Studies* 26, no. 1 (2011a): 14–24; Francesco Lapenta, "Locative Media and the Digital Visualisation of Space, Place and Information," *Visual Studies* 26, no. 1 (2011b): 1–2.
32. Eunjeong Choi, "KakaoTalk Mobile App Case Study," *Korea-Marketing* (2013), accessed September 10, 2013, www.korea-marketing.com/kakaotalk-mobile-app-case-study/.
33. Choi, "KakaoTalk."
34. Choi, "KakaoTalk."
35. James Russell, "After Making Money in Korea, Mobile Chat App Kakao Talk Takes Its Games Service Global," *The Next Web* (November 20, 2012), accessed July 5, 2013, http://thenextweb.com/asia/2012/11/20/after-making-money-in-korea-mobile-chat-app-kakao-talk-takes-its-games-service-global/.
36. Larissa Hjorth and Heewon Kim, "Being There and Being Here: Gendered Customising of Mobile 3G Practices Through a Case Study in Seoul," *Convergence* 11 (2005): 49–55.
37. Brian Sutton-Smith, *The Ambiguity of Play* (London: Routledge, 1997).

38. M. Hand, *Ubiquitous Photography* (Cambridge: Polity Press, 2012), 105.
39. Mizuko Ito and Daisuke Okabe, "Personal, Portable, Pedestrian Images," *receiver* 13, (2005), accessed July 5, 2013, www.receiver.vodafone.com/13/.
40. Ito and Okabe, "Personal, Portable, Pedestrian."
41. Hand, *Ubiquitous Photography,* 138.
42. Ito and Okabe, "Personal, Portable, Pedestrian."
43. Larissa Hjorth, *Mobile Media in the Asia-Pacific: Gender and the Art of Being Mobile* (London: Routledge, 2009).
44. Larissa Hjorth and Sun Sun Lim, "Mobile Intimacy in an Age of Affective Mobile Media," *Feminist Media Studies* 12, no. 4 (2012): 477–484.
45. Kenneth Gergen, "The Challenge of Absent Presence," *Perpetual Contact*, eds. James Katz and Mark Aakhus (Cambridge: Cambridge University Press, 2003), 227–41.
46. danah boyd, "Facebook's Privacy Trainwreck: Exposure, Invasion, and Social Convergence," *Convergence: The International Journal of Research into New Media Technologies* 14, 1 (2008): 13–20; danah boyd and Alice Marwick, "Social Privacy in Networked Publics: Teens' Attitudes, Practices, and Strategies" (2011), accessed July 5, 2013, http://papers.ssrn.com/sol3/papers.cfm?abstract_id=1925128; danah boyd, "Social Network Sites as Networked Publics: Affordances, Dynamics, and Implications," *A Networked Self: Identity, Community and Culture on Social Network Sites*, ed. Zizi Papacharissi (New York: Routledge, 2011), 39–58.
47. Kwang-Suk Lee, "Interrogating 'Digital Korea:' Mobile Phone Tracking and the Spatial Expansion of Labour Control," *Media International Australia* 141 (2011): 107–117.
48. Shin-Dong Kim, "The Shaping of New Politics in the Era of Mobile and Cyber Communication," *Mobile Democracy*, ed. Kristof Nyiri (Vienna: Passagen Verlag, 2003), 317–326.
49. Larissa Hjorth, "The Place of the Emplaced Mobile: A Case Study into Gendered Locative Media Practices," *Mobile Media & Communication Journal* 1, no. 1 (2013): 110–115.
50. boyd, "Social Network Sites as Networked Publics."
51. Lee, "Women's Creation of Camera Phone Culture."

3 Mobile Communication Technologies and Spatial Perception

Mapping London

Didem Özkul

Mobility and location matter a lot to us. Starting with one of our first spatial explorations, crawling, we learn to make sense of our geographical world on the move. Our spatial and social interactions help us identify not only our being-in-the-world but also places, constructed either as a result of those interactions or designed to host those interactions. We are not only observers in this geographical world but "are ourselves part of it, on the stage with other participants."[1] We continuously make spatial decisions that involve a continuous movement of our bodies, goods, and information, at all scales. These decisions range from how we get or send things from any given location to another, as well as the means and devices we use for those purposes, to how we deal with unexpected problems on our way, such as a train line delay, a punctured tire, or low reception on our mobile phone. The technologies we use to accomplish different types of mobilities play significant roles in how we make those decisions. Mobile communication technologies (as "miniaturised mobilities"[2]) make an especially important contribution to how we perceive and understand our spatial environment.

Scholarly works on mobile communication technologies have grappled with the effects of these technologies on our everyday life, focusing on changes in social and spatial practices of everyday life[3] and on the great extent to which these technologies blur the lines between public and private space, work and personal life, and social cohesion.[4] With the introduction of the GPS-enabled features of smartphones, the focus of mobile media research has shifted toward the analysis of location-based applications and their use in everyday life.[5] Although recent works explain locative media use in relation to theories of space and place[6] and question the changing definitions of location,[7] further empirical study is needed to explore how we use locational information to navigate in everyday life. Admittedly, many disciplines, such as environmental psychology and cognitive approaches to geography, have long grappled with the question of how we acquire such spatial knowledge and navigate in any given environment, as well as how we *locate* ourselves in any given space. However, the findings of these studies either have not been empirically incorporated into mobile and locative media research or have been taken for granted to explain certain phenomena theoretically.

In this chapter, I explore how locational information use on mobile communication devices changes spatial practices and navigation in London in relation to locations and places, which "in sum comprise his or her geographical world."[8] My focus is on the use of locational information both as external references and as sources of direct experiences in creating an experience of a city since "both sources of spatial information have to be combined in the cognitive map of an individual."[9] Hence, the concern is not what mobile technology users do but what they *experience* and how mobile maps could play a role in those individual experiences. My approach is neither cognitive (although I used sketch mapping as a tool to stimulate group discussion) nor behavioral. According to cognitive approaches in geography and urban planning, structuring and identifying any environment are innate abilities of human beings,[10] and we use internal and external references.[11] However, the way we do it is not a "mystic instinct"; "rather, there is a consistent use and organisation of definite sensory cues from the external environment."[12] Learning and making sense of spatial environments can rely on primary experiences, such as walking in a city, and on secondary sources, such as road signs and maps.[13]

Following the work of Seamon, I approach the relationship between spatial behavior and locational information use as a phenomenon. In contrast to the view of the cognitive theorists, Seamon argues that cognition plays only a *partial* role in everyday spatial behavior. In a similar vein, he also opposes the view of the behaviorist perspectives and argues that "the pre-reflective knowledge is not a chain of discrete, passive responses to external stimuli; rather, that the body holds within itself an active, intentional capacity which intimately 'knows' in its own special fashion the everyday practices."[14] Employing a similar approach, I blend cognitive, behavioral, and phenomenological approaches with empirical data from participant sketch maps of London and focus groups. Hence, this chapter provides a multidisciplinary approach to understand the spatial experiences of mobile technology users in London and how they "learn London step by step" through mobile and locative media.

To understand how users of mobile technologies experience any given urban space, I organized seven workshops in London in the year 2012.[15] In those workshops, following the method developed in Kevin Lynch's *The Image of the City*, I asked 38 participants in groups of five to eight to draw sketch maps of London and discuss their maps with each other in the form of a focus group. All of the group discussions were audio-recorded and transcribed. Using an artifact that participants themselves create and reflect on helps in explaining and understanding spatial relationships better than using verbal elicitation alone. When explaining things verbally, we also use spatial imagery and metaphors.[16] Thus, cognitive maps can also be used to create a context and content for social interactions, just as maps in general today "have changed from something that can spatialise social information to something that can socialise spatial information."[17] Drawing sketch maps encourages research participants to contribute and reflect on the unarticulated, the hidden, or the unsaid about their experiences of the city they live in and how they acquire that spatial knowledge over time. Throughout this chapter, participant maps and focus group conversations are used to support the argument that *mobile* maps are not only external or supplementary sources in creating an experience of a place but are also sources of direct experience and spatial participation.

Based on the analysis of empirical data, one of the major uses of locational information is related to navigation and map use in London. Under this category, spatial orientation includes basic forms of navigation such as walking, using public transportation, cycling, driving, as well as (different) uses of various maps (especially Google Maps on smartphones), both as primary and secondary sources of spatial learning and acquiring a sense of new places and as a means of direct experience of the spatial environment.

THE SHIFT FROM LANDMARKS TO SMARTPHONES IN ACQUIRING SPATIAL KNOWLEDGE

According to Kitchin and Blades, "primary learning is navigation-based, with the collection and processing of spatial information explicitly linked to an individual's interaction with an environment through spatial activity."[18] Several building blocks affect primary learning.[19] Environmental cues and features, such as landmarks and paths, as well as memorizing ordered views or scenes, are among the ways through which a cognitive map of the spatial environment can be formed.[20] According to behavioral approaches to habitual movement, there has to be an external stimulus that reinforces a particular pattern of spatial choice and behavior.[21] One might consult a smartphone application to find a restaurant nearby and the quickest route to get there. However, on the way to that restaurant, the same person could receive a photo as an attachment to a text message from a friend, showing the traffic jam on that route, and the recipient might therefore decide to take another route. In this scenario, what that hypothetical person does to find a place may not be as important as how she or he experiences the spatial environment through that technology. Hence, I contend that any mobile communication technology serves the processes both of obtaining a cognitive map and of stimulating a spatial behavior.

SET PATTERN OF DEVELOPMENT AND IMAGEABILITY

During the development of a cognitive map, there is a "set pattern of development."[22] Among the research participants, using landmarks and paths was a common method to navigate and to create a sense of place in London. They stressed the importance of landmarks, which were sometimes referred to as "checkpoints," "monuments," or "basics of London." In this respect, the River Thames, as a unique landmark, had a particular importance. It was the first thing drawn by most of the participants. Vicky told me why she started her map with the river:

> My first drawing was the River Thames. I see the river as a natural landscape and also it structures the whole city as south and north. It is quite unique. (Vicky)

Once it was placed on the sketch maps, participants started adding connections, revealing the "imageability"[23] of the city (see Figures 3.1a and 3.1b).

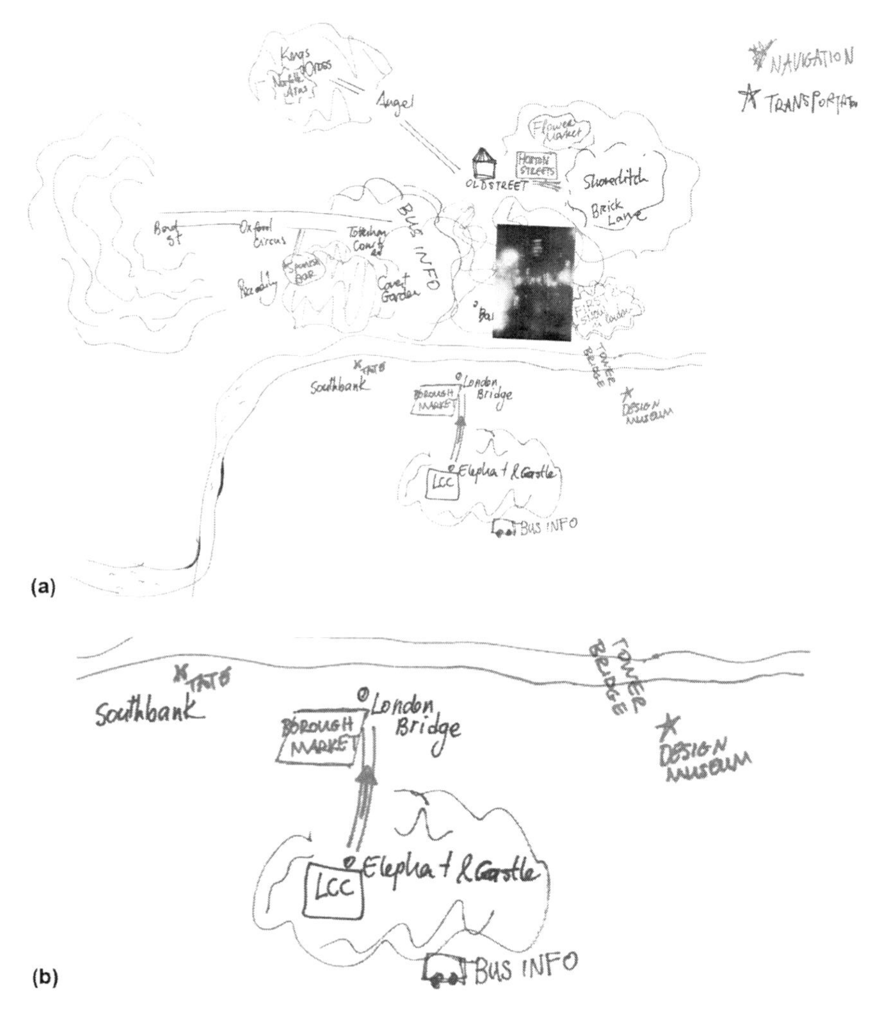

Figure 3.1 (a) Mary's sketch map of London. (b) "Design Museum and Tate are like two checkpoints for me to see what is where in the East" (Mary).

Siegel suggests that "an individual notes and remembers landmarks, and once landmarks are established an individual can 'attach' actions to these, so that the pattern of landmarks and actions is encoded as a route."[24] Walking in London, similarly, became the "attached action" for some respondents. Although it was apparent that the set pattern of development started with the Thames in many maps, the locations of other landmarks were altered.[25] Therefore, participants expressed some degree of difficulty in drawing and placing the landmarks on their maps, although they acknowledged that they knew the locations of them by heart once they started walking in the street and saw them:

> I don't know, I can map them actually when I am in the street. So instead of drawing them and placing them on a piece of paper. . . . It is quite difficult, but when I am walking . . . I have, like, checkpoints to know, in my head somewhere. (Mary)

Landmarks become especially important when one gets lost. Getting lost can create some sort of dependency on others' spatial knowledge or on way-finding tools such as maps. Anxiety resulting from not having the knowledge of one's whereabouts can provide the grounds for using smartphones. For instance, many respondents said that they did not look for street maps to find their way when they get lost or ask passers-by for directions. Rather, they preferred to place a mobile call or look up directions on their smartphones. By having the technology in hand, the dependency on (the spatial knowledge of) others can start to take a form of dependency on a particular technology: mobile maps, especially Google Maps on smartphones.

"MY MAP OF LONDON IS THE GOOGLE MAPS OF LONDON!"

A map can be defined as a "stable, combinable and transferable form of knowledge that is *portable* across space and time."[26] These pocket maps, as well as the mobile phone maps, "are each designed to be portable and to function while moving through space."[27] Among the research participants, Google Maps, either on the laptop or on the smartphone, is the most used type of map. Participants stated that they don't carry traditional paper maps anymore. Instead, they either use their smartphones, carry printouts of Google Maps, or *draw* sketch maps from the Google Maps. The ones who had a smartphone stated that they use the Google Maps application, while others, when they get lost, who only have a traditional mobile phone call or text people who they think might help them find their way, or they rely on others' smartphones:

> I don't remember the last time I got lost, but usually I end up calling someone whom I know would be by the computer. To look it up on the map for me. (Jackie)

> When I'm lost I usually text 5 people or so, sending the same text like "where is this place?" One of them would be by the computer. (Kier)

Instead of calling or texting someone once lost, some prefer to check Google Maps before leaving home. However, they usually draw sketch maps or take screen shots to carry with them.

> I do use a lot of Google Maps. What I do is, before I go somewhere, if I don't know the place, I draw a little map of that place and just take it with me. And sometimes, most of the time I forget it but then because I've drawn it then I remember it well. (Joanne)

Additionally, London's cosmopolitan nature and cultural differences can also play an important role in a shift from asking people directions to relying on mobile technologies, inasmuch as the commuting culture in London can sometimes prevent people from interacting with fellow commuters. Although a city is defined as "a human settlement in which strangers are likely to meet,"[28] one usually feigns ignorance to prevent social interaction with strangers, similar to Simmel's blasé attitude.[29] "This mental attitude of metropolitans toward one another we designate, from a formal point of view, as reserve."[30] This reserve was defined by some informants as the common culture in London, which they argued to be triggered by the use of mobile technologies as people are engaged with their mobile devices (smartphones, Kindles, iPads) while commuting. So it sometimes becomes impossible to create eye contact and ask a direction. On the other hand, asking for directions from fellow commuters is not always the best option because it is hard to tell who has local knowledge of London:

> There are so many people who don't know London. Non-Londoners. . . . So you don't know whom to ask beforehand. (Marianne)

In such circumstances, smartphones are seen as more reliable sources of locational information. Additionally, nowadays one is usually *expected* to carry such a mobile technology to navigate in London:

> One time I was with my best friend and she came to Waterloo, we wanted to walk to King's Cross and we wanted to cut through Pimlico. It was just a usual Saturday. Location-wise we just had to go straight ahead. If we took that road, you'd gonna be fine and we'd find it. But I wanted to be sure, because you know roads can be . . . So I stopped a guy, kind of a city boy more than a guy, kind of a city guy, just to ask if location wise are you on the right way . . . and he was like "don't you have a TomTom?" and I was just like "no, we don't." I wouldn't ask you if I did! (Sally)

> Once I asked somebody and they got their phone out and found the place for me. (Harriett)

No matter in what form, participants of the study stated that they use *mobile* maps. However, there is a significant difference when it comes to using maps on a smartphone. The map on a smartphone, initially a secondary source of data, becomes an interface that guides newcomers to learn how to navigate in London and that enables Londoners to *explore* different aspects of the city. As one of the participants commented on her sketch map of London at the end of the study: "My map of London is the Google Maps of London!" Thus she revealed a shift in the usage of maps in everyday life from maps as secondary sources of building spatial knowledge to maps as the source of direct experience. Hence, mobile maps are used not only for directions but also for different social and spatial explorations in a city.

FROM MAPS AS "SECONDARY SOURCES OF BUILDING SPATIAL KNOWLEDGE" TO MAPS AS "SOURCES OF DIRECT EXPERIENCE"

> At every instant, there is more than the eye can see, more than the ear can hear, a setting or a view waiting to be explored.[31]

The main distinction between relying on primary or secondary sources in acquiring a cognitive map is that the latter is a supplement to the former, which creates a direct experience.[32] Spatial information derived by direct experience is different from that acquired from maps.[33] "Maps show the spatial relationships between all the places *represented* on the map, but when an area is learnt from direct experience this knowledge has to be constructed gradually."[34] So maps, in this context, can only be supplements to direct experience of the environment. However, using mobile maps (especially Google Maps' Street View and/or 3D-view components) could actually lead to direct experiences of different places.

FAMILIARITY AND UNFAMILIARITY

Maps on smartphones provide the users with familiarity. Familiarity is a cognitive component of place attachment.[35] Hence, *mobile* maps can contribute to place attachment and establish a sense of a new place. According to Fullilove, "to be attached is to know and organize the details of the environment."[36] This type of locational information use can thus create a form of attachment to a specific spatial environment.

Kier, while explaining his use of maps in London, emphasized how Street View was different for him.

> You can actually see the whole place on Street View! Yes. You can see all the London landmarks, you can see all the landmarks from the world. Seeing it is very strange on a 3D virtual map. (Kier)

By making the landmarks of the world available to its user and by making it possible to walk in the virtually represented city, mobile maps can create for some a *similar* (if not exactly commensurate) experience of wandering in a city. This experience of similarity originates from the mobile maps feature of locating the user. Farman discusses that "the point of view offered by these maps engages the user along a spectrum from 'disembodied voyeur' to 'situated subject.'"[37] However, their success in helping users learn how to navigate in a city is still questionable according to this study because many participants reported that relying heavily on maps on mobile devices has a *side effect*, which some of the respondents worded as "*paying attention, but to where?*" For example, one of the respondents, Larry, told me that he uses Google Maps on his smartphone most of the time to navigate in London. For that reason, he believes that he does not *remember* the routes and directions.

> I usually go to places and forget about them until I see them again. (Larry)

Some of the respondents, such as Joanne, also pointed to such a problem in acquiring spatial knowledge by making the following analogy.

> It is a bit like a friend who knows the area so well that takes you to places. So when you're with someone who knows the area you just go with them. You don't even pay attention. (Joanne)

Similarly, Marianne also discussed why and how she does not pay attention to her surroundings while she is using Google Maps on her smartphone to navigate in London.

> If I don't have the technology with me, I think I'd remember more how to get to places. (Marianne)

Thus, although mobile maps can create a familiar sense of any given place, they also cause us to be unfamiliar with our usual, everyday spatial environment. Interestingly, despite the fact that many respondents argued that relying on mobile maps makes it harder for them to acquire spatial knowledge, they could not really classify it as a negative attribute. Instead, they expressed that their experiences of London became more enjoyable as a result of coincidental encounters and explorations through mobile maps.

SERENDIPITY: "WILL I FIND SOMETHING INTERESTING?"

In addition to using Google Maps on smartphones to find their way in London, some of the respondents explained that they use it also to search for and find places, which may go unnoticed in London's fast paced metropolitan lifestyle. By locating the user and then displaying nearby places, mobile maps can create serendipitous experiences, especially for participants such as Marianne, who enjoys discovering new places by coincidence:

> When you look at yourself on the map you notice that there's something by that you should check out. You wouldn't have had otherwise without that technology. (Marianne)

We sometimes want to get lost willingly in a city to discover new things and places. Mobile maps and location-based applications serve this desire of getting lost and exploring new things. Although the first thing that comes to mind when one thinks of maps is to *find* a place, one usually ignores the *different connotations* of finding a place, which can also be *exploring* and noticing something *different by chance*. I believe this is due to the categorization of maps as secondary sources of acquiring spatial information. However, the first maps were made to explore and find new routes and to discover new places all around the world. This forgotten dimension is reintroduced with location-based services, following the introduction of GPS to everyday life.

The will to find and experience something different also had a social aspect, with some of the participants indicating how they use locational information to discover new places and socialize with people in London. When participants in one of the sketch-mapping focus groups started talking about smartphone applications such as RunKeeper, which allows the user to share their routes with their networks, the social aspect of locational information sharing became clearer. What encourages some of the participants to share their random daily routes with fellow app users is the expectation for people to join in. This was also clear from the sketch map of one of the participants, Marianne, where she drew her running route along the Thames and talked about sharing her route with her friends (Figures 3.2a and 3.2b).

> I've got an app that did that, I used it once and I never used it again like 'who cares?'. I wouldn't look at anyone else's so I didn't. I don't see why you'd care about my route here. I felt like I was doing it to show off. (John)
>
> But people can join in. . .! (Marianne)

However, there was some sort of a coincidental use of the maps in smartphones. As some of the participants discussed among themselves and

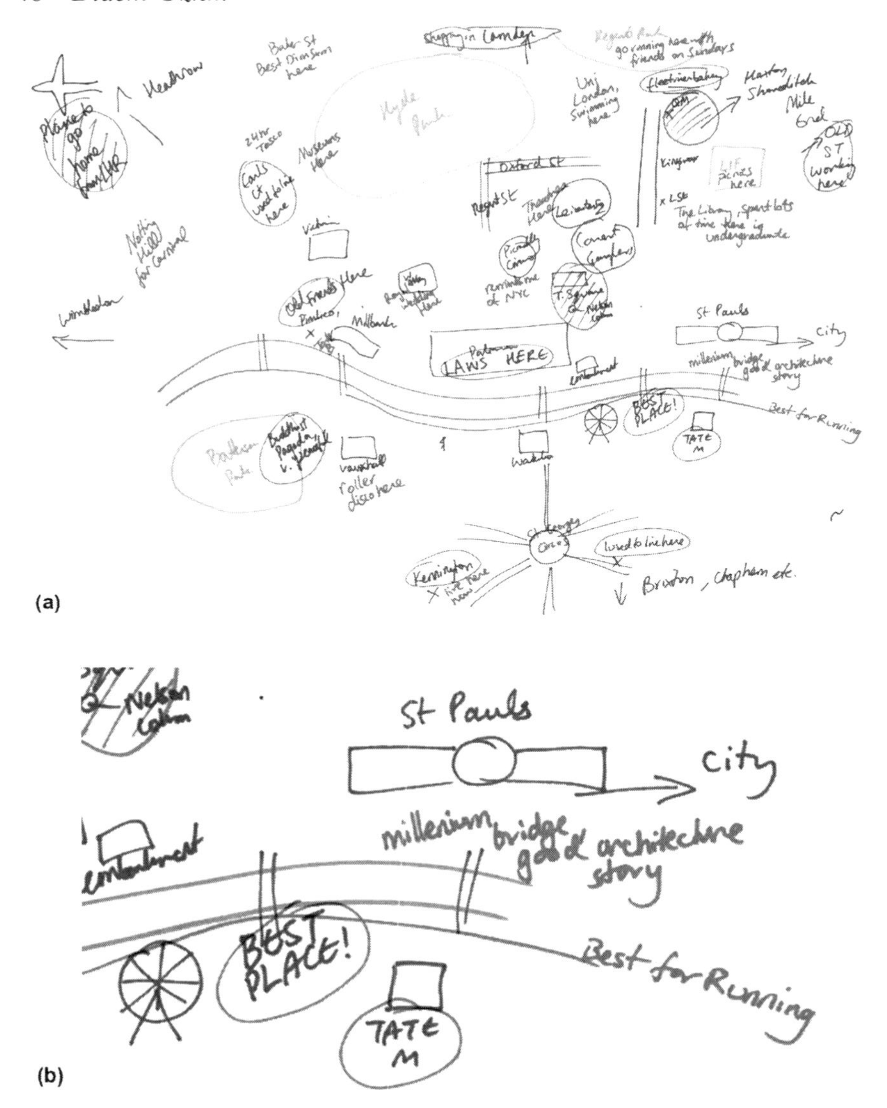

Figure 3.2 (a) Marianne's map of London. (b) Marianne's running route along the Thames.

explained, mobile maps, with the feature of locating the user, is also used to find something nearby. This type of use is mainly for "*micro-navigation*" as one of the respondents, Carlos, explained.

> I suppose, I just mainly use it for navigation, so . . . I don't know it is hard because I think I know London pretty well. I suppose I am using navigation to micro-navigate rather than broadly navigate. Because

> I know where to go if I want to go to Shoreditch or Chiswick or Fulham or wherever I know . . . I don't really have to look at my phone. It is probably more assuring to look at my phone rather than finding out where to go. But then when I arrive at somewhere, or if I am meeting at a specific place, it is more micro-navigation so that I know what specific street to go down. In my head, I know the tube map very well and the main layouts of roads and areas. (Carlos)

Exploring new things in a city is not only specific for users of those applications and those who share their locations. Nonusers who are somehow connected to those users via social networking can experience a *new* sense of a place based on their friends' locations. In addition to social use for sharing routes, one participant, Marianne, also revealed another aspect of such locational information sharing, which is keeping one's network up-to-date about the traffic.

> When I was living with my friends and we all went to LSE, and if there's lots of traffic and I'd take a photo of the traffic and send it to them so they would wake up earlier. So if I see something interesting, I'll take a photo of it and send it [to the classmates]. (Marianne)

On the other hand, the recipients of such locational information are not always users of such mobile technologies. For example, one of the respondents, Joanne, explained that although she did not use a smartphone or share locational information, she checks her friend's check-ins or geotagged photos on Facebook:

> On Twitter, or on Facebook for instance, if I see someone going somewhere then I look at it and if I think that it is nice I'll add it on my Facebook. It is basically the places where I want to . . . then I say 'Oh, I should go there' and I would go one day, definitely go there. That's how it works for me at least. (Joanne)

In either case, whether one owns a smartphone or not, it is apparent that in order to acquire a sense of (especially a new place in) London, users of mobile technologies started to rely on the mobile communication and location-aware features of those devices.

CONCLUSION

Maps, as tools for transferring knowledge or discovering and navigating in places, have always serviced our need and desire to be mobile. However, with the introduction of GPS-enabled mobile technologies into our everyday lives, users of these technologies have started to become dependent on the routes or directions they generate or recommend to us. Hence, they have the potential to affect how one navigates and acquires spatial knowledge. This potential effect is not only a result of having the technology on hand, but it also originates from the very nature of everyday life in big metropolises such as London. Having said that, landmarks still play a crucial role as primary sources in establishing spatial knowledge. As this study also reveals, maps on our mobile devices can be said to remind us of our own very exploratory and adventurous nature by allowing us to discover and explore things serendipitously, along with making us familiar with new places. However, they also have the potential to limit our spatial and social interactions if they are used *only* as secondary sources of spatial information. To conclude, most importantly, mobile and locative media do not act just as supplementary sources, but they can also be used as *sources* of direct experience, enhancing one's awareness of the spatial environment socially.

NOTES

1. Kevin Lynch, *The Image of the City* (Cambridge, MA: MIT Press, 1960), 2.
2. Anthony Elliott and John Urry, *Mobile Lives* (New York: Routledge, 2010), 28.
3. An indicative and certainly not exhaustive list includes Manuel Castells, Mireia Fernández-Ardèvol, Jack Linchuan Qui, and Araba Sey, *Mobile Communication and Society: A Global Perspective* (Cambridge, MA: MIT Press, 2007); Gerard Goggin and Larissa Hjorth, eds., *Mobile Technologies: From Telecommunications to Media* (New York: Routledge, 2009); James E. Katz, *Magic in the Air: Mobile Communication and the Transformations of Social Life* (New Brunswick, NJ: Transaction Publishers, 2006); Rich Ling and Scott W. Campbell, eds., *The Reconstruction of Space and Time: Mobile Communication Practices* (New Brunswick, NJ: Transaction Publishers, 2009).
4. See also Nicola Green and Leslie Haddon, *Mobile Communications: An Introduction to New Media* (Oxford: Berg, 2009); Gerard Goggin, *Cell Phone Culture: Mobile Technology in Everyday Life* (New York: Routledge, 2006); Maren Hartmann, Patrick Rossler, and Joachim R. Höflich, eds., *After the Mobile Phone? Social Changes and the Development of Mobile Communication* (Berlin: Frank and Timme GmbH, 2008); Rich Ling, *New Tech, New Ties: How Mobile Communication Is Reshaping Social Cohesion* (Cambridge, MA: MIT Press, 2008).
5. An exemplary list includes Adriana de Souza e Silva and Daniel Sutko, eds., *Digital Cityscapes: Merging Digital and Urban Playscapes* (New York: Peter Lang, 2009); Lee Humphreys and Tony Liao, "Mobile Geotagging: Reexamining Our Interactions with Urban Space," *Journal of Computer-Mediated Communication* 16, no. 3 (2011), 407–423.
6. Eric Gordon and Adriana de Souza e Silva, *Net Locality: Why Location Matters in a Networked World* (London: Wiley-Blackwell, 2011); Shaun Moores, *Media, Place and Mobility* (Hampshire, UK: Palgrave Macmillan, 2012); Rowan Wilken and Gerard Goggin, eds., *Mobile Technology and Place* (New York: Routledge, 2012).
7. Adriana de Souza e Silva and Jordan Frith, *Mobile Interfaces in Public Spaces: Locational Privacy, Control, and Urban Sociability* (New York: Routledge, 2012).

8. David Seamon, *A Geography of the Lifeworld: Movement, Rest, Encounter* (London: Croom Helm, 1979), 15.
9. Rob Kitchin and Mark Blades, *The Cognition of Geographic Space* (London: I. B. Tauris & Co Ltd., 2002), 47.
10. An indicative but not exhaustive list includes Roger M. Downs and David Stea, *Maps in Minds: Reflections on Cognitive Mapping* (New York: Harper & Row, 1977); Kevin Lynch, *The Image of the City* (Cambridge, MA: MIT Press, 1960); Yi Fu Tuan, *Space and Place: The Perspective of Experience* (Minneapolis: University of Minnesota Press, 1977).
11. Kitchin and Blades, *The Cognition of Geographic Space.*
12. Lynch, *The Image of the City.*
13. Lynch, *The Image of the City.*
14. Seamon, *A Geography of the Lifeworld: Movement, Rest, Encounter,* 35.
15. The project is part of Didem Özkul's PhD research, based at the University of Westminster.
16. Downs and Stea, *Maps in Minds: Reflections on Cognitive Mapping,* 12–27.
17. Gordon and de Souza e Silva, *Net Locality: Why Location Matters in a Networked World,* 28.
18. Kitchin and Blades, *The Cognition of Geographic Space,* 35.
19. Kitchin and Blades, *The Cognition of Geographic Space,* 35.
20. Kitchin and Blades, *The Cognition of Geographic Space,* 35.
21. Seamon, *A Geography of the Lifeworld: Movement, Rest, Encounter.*
22. Kitchin and Blades, *The Cognition of Geographic Space.*
23. Lynch, *The Image of the City.*
24. Siegel as cited in Kitchin and Blades, *The Cognition of Geographic Space,* 35–37.
25. Cognitive maps and hence sketch maps that represent those cognitive maps are not geographically accurate; see Kitchin and Blades, *The Cognition of Geographic Space,* 57.
26. Rob Kitchin, Chris Perkins, and Martin Dodge, "Thinking About Maps," in *Rethinking Maps: New Frontiers in Cartographic Theory,* eds. Martin Dodge, Rob Kitchin, and Chris Perkins (New York: Routledge, 2009), 15.
27. Jason Farman, *Mobile Interface Theory: Embodied Space and Locative Media* (New York: Routledge, 2012), 46.
28. Richard Sennett, *The Fall of Public Man* (New York: Norton, 1974), 39.
29. Georg Simmel, "The Metropolis and Mental Life," in *Classic Essays on the Culture of Cities*, ed. Richard Sennett (New York: Meredith Corporation, 1969), 47–69.
30. Simmel, "The Metropolis and Mental Life": 53.
31. Lynch, *The Image of the City*, 1.
32. Lynch, *The Image of the City*, 1.
33. Kitchin and Blades, *The Cognition of Geographic Space,* 45.
34. Kitchin and Blades, *The Cognition of Geographic Space,* 45 (emphasis added).
35. Leila Scannell and Robert Gifford, "Defining Place Attachment: A Tripartite Organizing Framework,"*Journal of Environmental Psychology* 30, no. 1 (2010): 1–10.
36. Fullilove, as cited in Scannell and Gifford, "Defining Place Attachment," 3.
37. Farman, *Mobile Interface Theory,* 45–46.

4 The Social Media Life of Public Spaces

Reading Places Through the Lens of Geotagged Data

Raz Schwartz and Nadav Hochman

INTRODUCTION

How do we "know" a place? What are the modalities by which we can study the character of a social place today, examine social interactions within it, and trace its unique rhythm? The vast growth in geolocated and time-stamped social media data such as tweets, check-ins, and images promises to bring new challenges and opportunities in investigating these questions. This research is about city spaces and locative social media—what we can learn, what we cannot learn about these spaces, and what the practical lessons and applications may be. It is research that is a by-product of data-driven observation.

Public spaces offer us a way to escape the fast pace of urban life, they are places where people from various backgrounds can gather and participate in community meetings, events, and festivals. They serve as landmarks, points of reference, and as assets for local communities. But what do we know about these places, and how can we study them? Until recently, researchers who wanted to study the social interactions and usage of specific places had a very limited set of tools (such as surveys, documentation via photos or videos, and crowd counting) to achieve their goals. In contrast to these earlier forms of observation, the high granularity of social media data provides researchers with the ability to visualize, analyze, and conceptualize the structure and character of particular geographical areas according to the multitude of social media activities going on within them.

As mobile devices become increasingly ubiquitous and people are constantly encouraged to document their everyday activities, researchers, planners, and local community organizations are faced with new opportunities to better understand their locality through access to publicly shared social data. In this way, data generated by users of locative media tools might provide new ways to study activities in specific places and uncover local insights.

These insights, which are usually available to only a limited group of people such as community leaders and local advocates, were always highly coveted. Over the years, social scientists and urban planners used methods like interviews, surveys, people counting, or photo and video documentation to study particular places and the social participation that takes place there.[1] The famous American urbanist William Whyte, for example, published in 1980 one of the first works that tried to study and compare various public places in New York City.[2] Whyte set out to examine what makes a certain public plaza more sociable than another, and for that he used video recording, crowd mapping, and interviews to synthesize several recommendations. Georges Perec conducted similar work on a much smaller scale. During a three-day period, Perec documented the nonevents in the mundane life of Place Saint-Sulpice, a public plaza in Paris as he observed it through the window of a nearby coffee shop.[3] More recently, Joachim Höflich followed the movement of people using mobile phones while walking through an Italian public plaza and drew maps that uncovered their behavioral patterns.[4]

Although these studies provide meaningful insights about how people interact with one another as well as with the place itself, they are limited in scope and can provide only a glimpse into a specific time period in a confined geographical area. In this way, comparing the use and activity of specific places turns out to be a challenging task that requires vast resources and considerable time. Computer scientists who attended this problem developed various techniques that utilized social data on a large scale from many places, people, and times. However, these studies rarely examined the particularity of specific places within the city. Favoring an aggregated image of the entire city or of confined regions within the city, these representations override the unique sociocultural aspects of a place and its dynamic structure in space and time.

Following Whyte, 30 years later, a study traced three public parks in New York City, this time utilizing geotagged social media data from locative media platforms such as Instagram and Yelp. In the following sections, we will survey existing literature and showcase our data-driven exploration of these public spaces. We first illustrate several techniques that can help researchers approach the study of social media data from public places. More specifically, we use computational methods to look at various factors that can help us understand people's use of and navigation through three public parks in New York City. The results of this analysis show differences in park usage based on time of day, the prominence of public art as it appears in this data, as well as unique activities and interests (based on textual analysis of references and captions from Yelp and Instagram data) that are offered in each of these public venues.

To summarize, we offer a preliminary conceptual and analytical work for a more nuanced understanding of physical places based on the social media data generated within them. By applying computational tools to data sets from various location-based social networks, we are able to identify special characteristics and produce a comparative analysis of these places as well as examine the opportunities of using geotagged data for the study of urban public spaces. This research sets out to depict what "the social media image" of a public place is, as well as how we can use social media data to compare these places and trace their differences? And, finally, we consider who is left out? What demographics, activities, and social interactions are not captured by this data and should be taken in to consideration when using this kind of data in decision-making processes?

THE STUDY OF URBAN SPACES

The social scientific study of spaces via social media data can be tracked back to first explorations and usage of aerial photography in the aftermath of World War II by social scientists, who attempted to mash together for the first time aerial and ground/street level perspectives for the study of the social, economic, political, and cultural aspects of geographic areas. The researchers combined emerging forms of militaristic aerial representations with ethnographic fieldwork to "reveal" the structure of a particular geographic area, or what they called a "social space," that is, "the division of a space according to the particular norms of a group." As explained by Jeanne Haffner, this new representational space was abstracted "from the chaos of the ground but not divorced from it"; it was "a model of the relationship between these analytical categories—a spatialization of complex social and economic relationships within a particular urban environment."[5]

With the growth in geolocated social media data, these urban representations take new forms and shapes, created by aggregating hundreds of millions of peoples' tweets, check-ins, images, and videos into condensed representations of neighborhoods, of cities, and of the entire earth. One trend in social media research has focused on identifying landmarks and points of interest that have high visibility (the most shared or tagged) in social media data.[6] Other studies utilized users' profile data such as home city and past activity to denote a place most frequented by tourists or locals.[7] The resultant representations, however, completely neglect the temporal element of the data, create an aggregated, cumulative, and constant activity profile of an area, and do not capture real-life social dynamics.

Another strand in social media research aims to uncover collective geographic patterns in confined areas. For example, many studies have examined mobility patterns and urban dynamics. More closely related to our interest here is work conducted by Noulas and colleagues, who have examined geographical clustering based on the aggregated activities in an area, while Cranshaw and associates and Zhang and associates offer a glance into the areas of the city that like-minded people visit.[8] The Livehoods project, for example, shows how a series of bars, restaurants, parks, or shops may carry a strong connection with a certain group of people who include them in their social media profiles.[9]

Although these studies, overall, continue to ignore temporal dynamics in the data and geographic area is considered in aggregate, they also apply clustering or homogeneity models in order to extract social *similarity* between different geographical locations in the city. In this way, these models offer to restructure the city by means of tracing its social boundaries (where particular social groups go to them and when) based on users' social media activities, as opposed to predetermined values or traditional hierarchies (i.e., municipal borders). The results, however, are homogeneous clusters of fixed entities that erase the *particularity* of a singular place and its dynamic nature in favor of its aggregation and categorization with other similar types of places with different levels of abstraction (i.e., areas frequented by locals versus tourists). In other words, they override the particularity of a place in favor of its representation as a connected, networked whole with other similar places.

As information technology has evolved, mobile media has transformed the ways in which we understand and engage with places.[10] Existing computational research projects on the particularity of a place over social media networks are still rare in kind. A related and exceptional study in our regard examined the profile pictures of bar patrons in Austin, Texas, and illustrated how users' Foursquare profile photos conveyed information about the character and connotations of specific venues based on impressions about the types of people who frequent these venues.[11] In addition, Hochman and Manovich examined the dynamic representation of political, social, and cultural events in particular locations in the city.[12] Moreover, there is social science research examining the use of mobile phones and social interactions in public spaces. For example, Hampton and colleagues studied the impact of wireless Internet use on public spaces and showed that online activities in public spaces contribute to broader participation in the public sphere.[13] Recent studies have examined the use of location-based communication technology in public spaces, showing how the use of Foursquare changes and impacts people's sense of place and promotes parochialization and personal attachment to the public space.[14]

Users of these locative media platforms generate ongoing flows of information that provide a glimpse into the daily activity of city dwellers. For this work, we have opted to focus on two main platforms: Instagram and Yelp. Although the affordances of each of these platforms differ, the locative aspect of the data is at their core. Using the platforms' API (application programming interface) access, as well as a web crawler that was developed especially for this work, we were able to access vast amounts of data that were tagged to various public spaces in the city of New York.

The first of these, Instagram, is a photo- and video–sharing social network for the mobile phone that was first introduced in October of 2010. Instagram users can take pictures and videos, apply various filters, and share them with their friends on the application or in many other social networks. The popular service has over 150 million active users as of September 2013.[15]

When sharing a photo or video, users can choose to add it to their Photo Map, a personal visualization that showcases the users' activity based on geographic data. In this way, a photo that is added to this personal map includes the specific latitude and longitude of the place where the photo was uploaded. The location can then be displayed to other users in a timeline or news feed where participants can view photos taken by the users they follow.

In contrast to the visual-centric Instagram, Yelp is an online review community that started in 2004 as a website meant to provide its users an online urban guide. From a modest start as an e-mail service for local business recommendations, the platform has grown considerably and was later transformed into a social networking service that receives almost half of its traffic from mobile devices. In September 2013, Yelp had a monthly average of 117 million unique visitors, and its mobile application was used on 11.2 million unique mobile devices on average every month.[16]

In the mobile phone versions, users are encouraged to check in to a location and leave a review. Reviews on Yelp tend to be long and also include data such as the user's hometown and a venue score, information that is not available in other check-in services. Due to the relatively long life span of this service, it has a vast data set of reviews generated by its users. By the end of the third quarter of 2013, Yelp users (referred to as Yelpers) had submitted more than 47 million local reviews. Reviewers can also upload photos and mark other reviews as "Useful," "Funny," or "Cool."

Locative social media data can provide new perspectives to better understand the spatial aspects of a public space. As users publicly share check-ins, reviews, photos, and videos, we must ask ourselves what happens when we aggregate this data from several parks? What can this data tell us about the spatial conditions, points of interests, and groupings of people throughout these public spaces? In the following sections, we will examine the type of insights this data yields as well as their limitations.

SPATIAL READING

Each Instagram photo contains rich metadata. This information includes the username of its creator, the exact time the photo was taken, as well as the exact location from which it was uploaded. This location is annotated by latitude and longitude information that, when cross-referenced, creates a unique dot on the world's map. We started our exploration by retrieving all the publicly shared Instagram photos from three parks in Manhattan: Madison Square Park, Bryant Park, and Union Square Park. Each of these parks is located next to various office buildings, food venders, and public transport hubs. Using the Instagram API interface, we gathered all the publicly shared geotagged photos from these parks over a period of 6 months (November 20, 2012 to April 15, 2013). We ended up with the following numbers of photos: 14,593 for Union Square, 18,352 for Bryant Park, and 14,989 for Madison Square Park.

When plotting our data over a city map, we can see several interesting formations of patterns among the photos. For example, in the Union Square plot we can see how there are two main high-density locations: the middle south part and the top west part. In Bryant Park, on the other hand, there is only one center area along the middle part of the park that shows high numbers of photos. Finally, in Madison Square Park, we can see that a large portion of the photos taken at the park were tagged to the southwest part of it. When looking at the photos from each of these locations, we examine the reasons for these dense plots.

In Bryant Park, the center area of the park (or the great lawn) is the home of the ice skating rink during the winter as well as other community events during the year, such as free public yoga sessions (Figure 4.1b). The park, which was redesigned in 1930s by Robert Moses, the famous master builder of New York City, follows traditional symmetrical European architecture, including a center fountain feature and a main lawn area. This layout promotes the grouping of photos in its center, and, as we can see, the symmetrical plots of photos follows closely the traditional architectural layout of the park.

The plots of Union Square, on the other hand, display two different concentrations (Figure 4.1a). One concentration is located in the northwest part of the square and another one in its southern central part. When examining the photos from these two areas, we discovered that the northern grouping is attributed to the farmers' market that takes place on Mondays, Wednesdays, Fridays, and Saturdays, and the southern part serves as a public venue for demonstrations and public gatherings.

Madison Square Park (Figure 4.1c) has a relatively equal coverage throughout the park, except for the bottom west part, which consisted

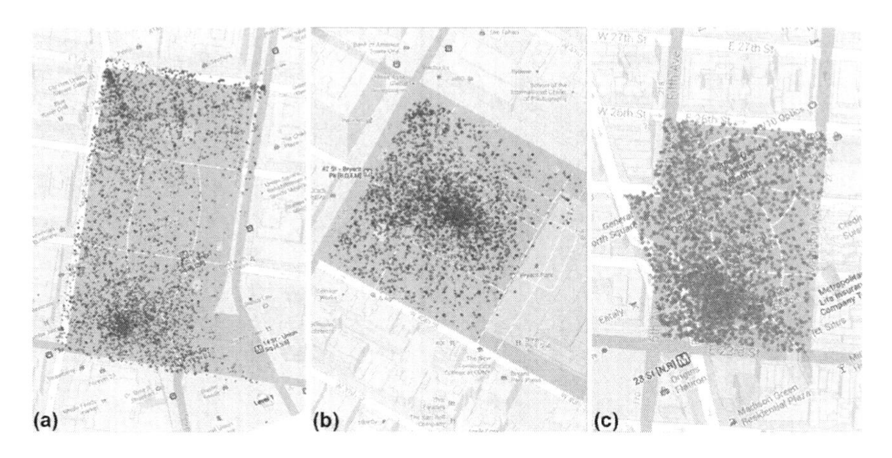

Figure 4.1 Geographical plotting of publicly shared geotagged Instagram photos November 20, 2012–April 15, 2013. Left: Union Square—14,593 photos, Center: Bryant Park—18,352 photos, Right: Madison Square Park—14,989 photos.

mainly of pictures of the Flat Iron building from across the street. This spread of the photos can be attributed mostly to the renovation of the park that was completed in June 2001. This renovation created several different points of interest throughout the park, such as a dog run, a center water feature, and a kids' playground; the renovation also brought in an ongoing series of outdoor sculpture exhibitions.

TEMPORAL READING

This data can also provide a more historical perspective and help us examine temporal trends and use cases for the parks over several months or years. In this way, we can study upload times of photos as an indicator of user activity in the park and then look at a time series and find out which days are busier than others. For example, we can plot the average rate of photos per hour in each of these locations (Figure 4.2a). Looking at these charts, we can see that most of the parks are busier around the afternoons, with Madison Square Park peaking at 4:00 p.m., Union Square at 5:00 p.m.,

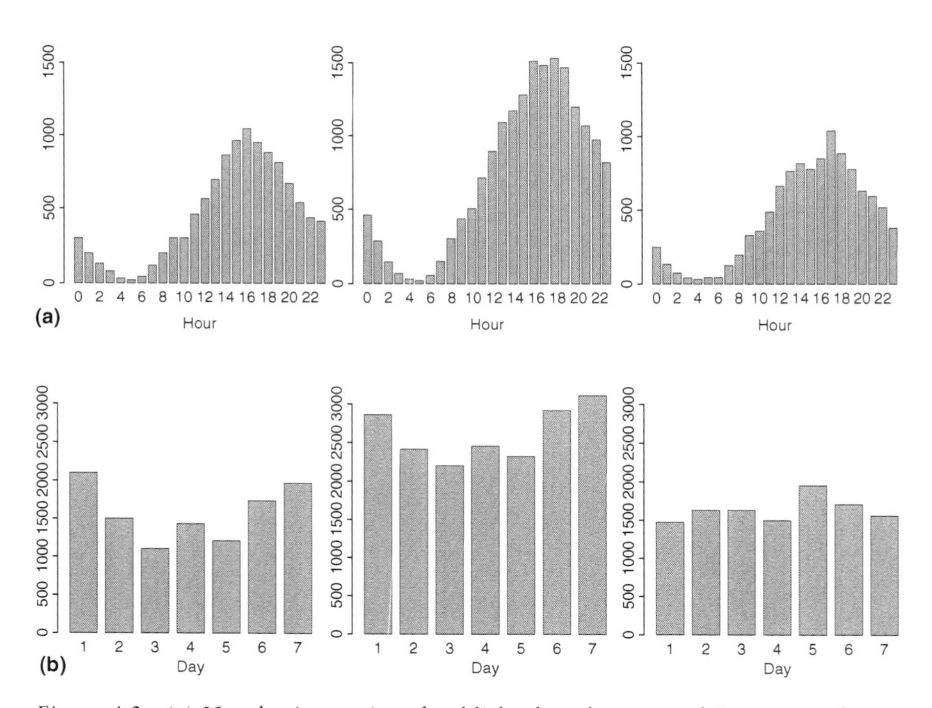

Figure 4.2 (a) Hourly time series of publicly shared geotagged Instagram photos November 20, 2012–April 15, 2013; Left: Union Square—14,593 photos, Center: Bryant Park—18,352 photos, Right: Madison Square Park—14,989 photos. (b) Daily time series plotting of publicly shared geotagged Instagram photos November 20, 2012–April 15, 2013 (number of photos per park as before).

and Bryant Park at 6:00 p.m. Even though the peaks are just about similar among these places, we can see how the number of photos during the afternoons and evenings at Bryant Park is a rate that is one and a half times higher than during the lunch hours (1,000 versus 1,500 photos per hour in the afternoons).

We can also extract daily patterns out of this data. In Figure 4.2b, we can see how photo-sharing activity is different in city parks throughout the week. Bryant Park, for example, shows a high number of photos during the weekend with its peak on Saturday, whereas Union Square peaks on Sunday and Madison Square Park on Thursday.

By looking at a particular physical location from a temporal perspective, we can uncover patterns that were not visible before. For example, we can see the effect of seasons, weather conditions, or catastrophic events, such as hurricanes on these places. When aggregating these temporal signals, we can start reading the rhythm of a place,[17] the cycles of visits throughout weekdays and during the hours of each day.

TRACING EXPERIENCES

In his work, Whyte looked at the different factors that make a successful public plaza. Among these were the availability and type of sitting, the configuration of the place in relation to the sun and the wind as well as the importance of trees, water, and art to the experiences of their visitors. Moreover, special attention was given to the existence of food vendors and restaurants as places that patrons can enjoy while exploring and socializing in the public realm. These factors play a large role in the experiences of people in these public spaces as well as in the social interactions that take place there.

Social media can provide us with a way to tap the visitors' experiences by following their online social interactions and activity. In this case, we looked at the reviews that were posted to these venues. By crawling the Yelp reviews that were tagged to these parks and using textual analysis techniques, we were able to extract the main themes that appear in these user-generated texts. In the word clouds in Figure 4.3a–c, we can see the occurrence of various words throughout the reviews. These visualizations display the main topics that people find interesting enough to share and review.

First, we can see the high occurrence of the word "people" in each of these word clouds (Figure 4.3a–c). Although it might not be surprising that this is among the most common word in these reviews, it does show how these places are centers for the social gatherings of many individuals. However, if we dig deeper and look at another form of visualization that shows relations between words, we can see how the word "people" is closely related to the verb "watch," displaying the special practice of "people watching" that takes place in these public parks.

This practice is also closely related to the topic of sitting conventions in these parks. Sitting has been much debated in the realm of urban

Figure 4.3 Word cloud plotting of Yelp review: (a) from Union Square—226 reviews; (b) from Bryant park—462 reviews; (c) from Madison Square Park—221 reviews.

planning, and designers have tried many different methods to provide sitting options that will be both sustainable and easily maintainable, as well as comfortable and usable for the general public. One of the recommendations in Whyte's work described the importance of movable chairs and tables.

This recommendation was applied in each of these parks, but the number of chairs and tables in Byrant Park exceeds by a large number those in the other parks. In 1988, Bryant Park was closed for a four-year project that enhanced the very traditional French garden design, improved and repaired paths and lighting, and placed movable chairs and tables throughout the park.[18] This difference among the parks can be also seen in the mentions of chairs and tables in the review. Although all of the parks have mentions of these, Bryant Park's visualization shows the magnitude of these occurrences together with their close relationship with the mentions of the verb "enjoy" (Figure 4.3b).

Food plays another important part in the experience of public parks and plazas. When examining the textual analysis results from these reviews, we can see how Shake Shack, a popular American fast-food restaurant, appears prominently in the Madison Square Park visualization. This is mostly due to the 2004 addition of this permanent food stand, which serves hamburgers, hot dogs, and shakes (Figures 4.4a, 4.4b. 4.4c).

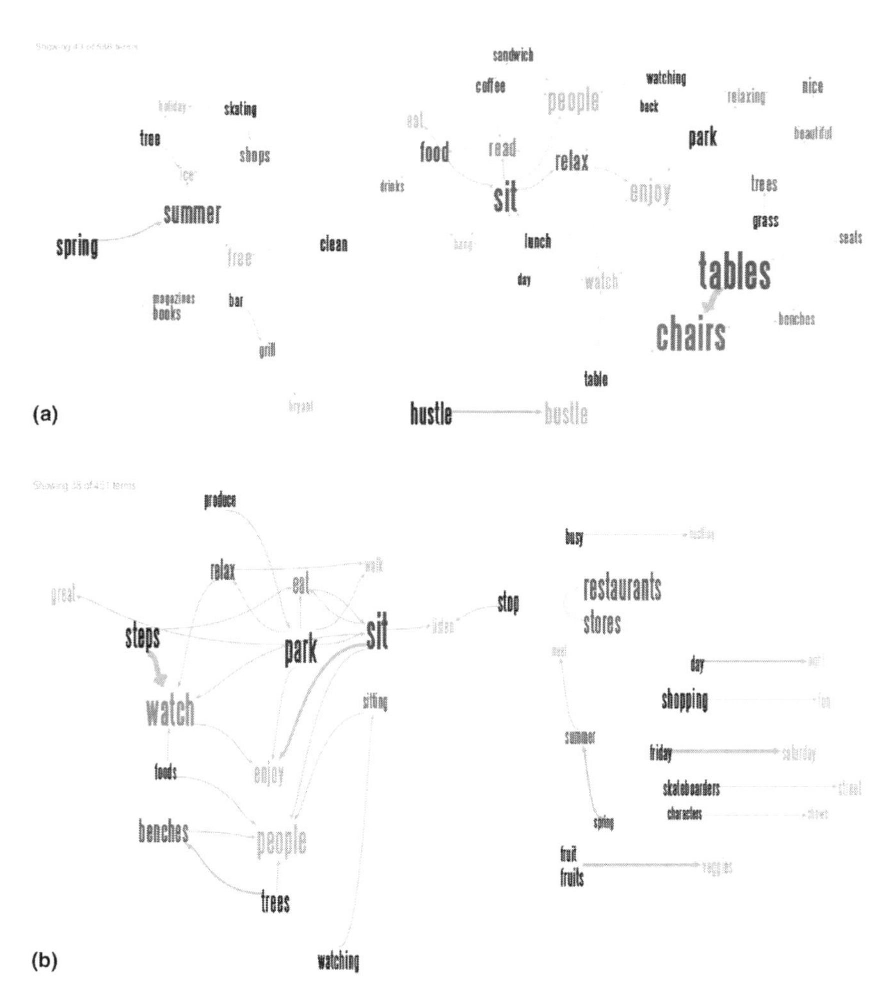

Figure 4.4 Phrase Net visualization displays networks of related words of Yelp Review: (a) from Union Square—226 reviews; (b) from Bryant park—462 reviews; (c) from Madison Square Park—221 reviews.

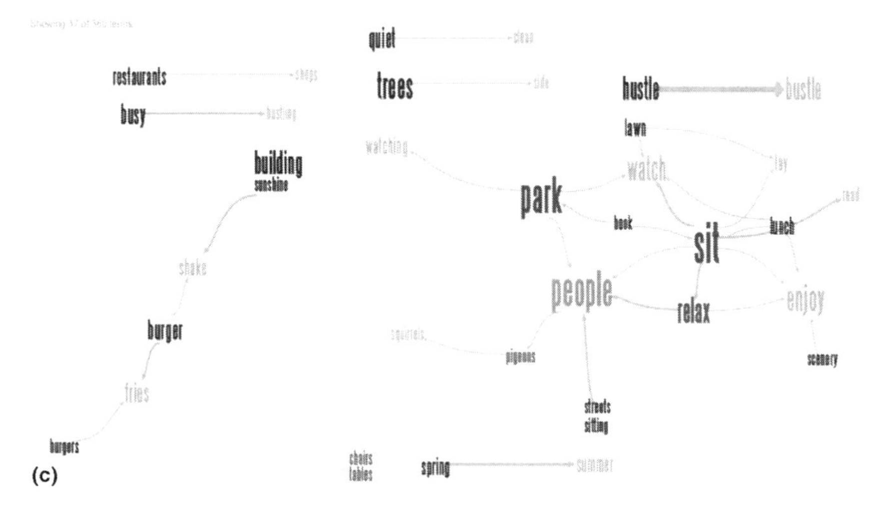

Figure 4.4 (Continued)

In addition, special features in the park can also come across in this kind of data. For example, the placement of art, a dog run, a farmers' market, or a skating rink are all activities and facilities that have an important part in the life of the public space (Figures 4.3a and 4.3c).

WHO IS LEFT OUT?

Contrary to other methods, such as small-scale observations or the various interviews and surveys that tried to capture the varied demographics in these publics spaces, social media data is prefiltered to a very specific group of users. While studying this data, we must take into consideration people who do not own a mobile phone or a smartphone. Despite the proliferation of smartphone usage in the United States, only 50 percent of mobile users have one. Moreover, only 30 percent of them use the location indication of their phone over social media.[19]

We are therefore left with a population who are reasonably tech savvy with the means to own and maintain a smartphone, who document what they do, who participate on social media platforms, and who share their daily activities publicly. This bias is particularly important when studying public spaces. As opposed to other places such as home and work, public spaces, also referred as third places,[20] are supposed to be places where people from various backgrounds can come together—a public sphere that welcomes people from all levels of society. When we use only geotagged social media data, we are inclined to ignore a large portion of the city's population.

Considering this kind of data as a detailed reflection of real life is problematic because it is far from being a precise indication of the location of social interactions. As Larissa Hjorth notes, many playful, creative practices emerge on these platforms,[21] and, as Barkhuus and colleagues have shown, expressing one's location over a social network may also signal mood, lifestyle, or life events and maintain or support intimate social relationships.[22] The manner in which certain locations or activities are named, captioned, or annotated can be understood as performative. Cramer and associates provide evidence for this claim in their study of Foursquare users and the common practice of creating "imaginary" places, as well as fictitious or creative names for places or events.[23] Humphreys also examines how practices of cataloging and archiving personal mobility and presence within a place encourage intimate bonding with friends and are used in the service of bragging or "showing off," self-promotion, making inside jokes, recording places as a memory aid, or receiving points or rewards for particular habits or actions.[24]

Moreover, in many instances, social media signals in public spaces are biased toward special events or activities that are out of the usual. Consequently, regular patterns are less well documented. For example, people who frequent the park in order to read a book, walk their dogs, or eat lunch will not necessarily be inclined to share this mundane activity on social media. And yet if a street artist or a band plays in the park, these instances might trigger more people to share their activity with their online friends and followers.

CONCLUSION

There is no denying that our perspectives and understandings of human activity in urban areas are changing. As a rapidly growing number of people constantly log and publicly share their daily activity, more and more companies, urban planners, and researchers are gaining access to highly granulated social data of urban activity that was not available before.

Studying a specific geographic place has always been a challenging task. Public spaces are therefore fertile ground to study geotagged social media data because they provide a place (if things are done right) that is designed to be an open, inviting public sphere that supports local social interactions. As Lofland notes, these spaces level differences between people that derive from their social circles and diverse backgrounds. It is the public sphere in its glory, a place to meet both friends and strangers.[25]

In this chapter, we have provided a glimpse into the type of data and insights that we can extract from studying these kinds of information flows, as well as the data that we cannot access, the demographics that are left behind, and the performance aspects of the data. It is therefore highly crucial to understand the biases and caveats that are inherited in this type of data. This data is biased not only by the specific demographics that generate it but also by the social practices and norms that these platforms enable. In this way, ideas regarding access to technology and online connectivity are joining ideas such as gaming and exhibition and performance of the self to tell us that many of the data points should not be considered as precise indications of location or social interactions.[26] They are inherently biased, and they should be treated as such.

To conclude, geotagged social media offer a new lens through which to better understand the use of urban public places. Every day, an ongoing stream of information is produced by various users who act in a constant documentation effort of the life of the park—tweeting, reviewing, checking in, and taking pictures of themselves or others (using the hashtag #peoplewatching), social events, and art installations. However, as with any other skewed lens, some artifacts receive more attention than others. It is an additional layer of information that can shed new light on old questions or give rise to new questions that were not available before. However, using this kind of data does not eliminate the need to perform additional, more traditional research, such as observations, surveys, and interviews. Only by applying a mix of both qualitative and quantitative methods can we advance our understanding of our public spaces in our effort to make them a better venue to socialize and enjoy as part of the experience of living in highly dense urban habitats.

NOTES

1. Erving Goffman, *Behavior in Public Places: Notes on the Social Organization of Gatherings* (New York: Free of Glencoe, 1963).
2. William Hollingsworth Whyte, *The Social Life of Small Urban Spaces* (Washington, DC: Conservation Foundation, 1980).
3. Georges Perec, *An Attempt at Exhausting a Place in Paris*, trans. Marc Lowenthal (Cambridge, MA: Wakefield Press, 2010).
4. Joachim R. Höflich, "A Certain Sense of Place: Mobile Communication and Local Orientation," in *A Sense of Place: The Global and the Local in Mobile Communication*, ed. Kristóf Nyíri (Vienna: Passagen Verlag, 2005), 159–168.
5. Jeanne Haffner, *The View from Above: The Science of Social Space* (Cambridge, MA: MIT Press, 2013), 82.
6. Lyndon Kennedy and Mor Naaman, "Generating Diverse and Representative Image Search Results for Landmarks" (paper presented at the Seventeenth International World Wide Web Conference, April 2008, Beijing, China).
7. Eric Fischer, "The Geotaggers' World Atlas," *Flickr*, accessed September 25, 2013, www.flickr.com/photos/walkingsf/sets/72157623971287575/.
8. Anastasios Noulas, Salvatore Scellato, Renaud Lambiotte, Massimiliano Pontil, and Cecilia Mascolo, "A Tale of Many Cities: Universal Patterns in Human Urban Mobility,"*PLoS ONE* 7, no. 5 (2012), e37027, doi: 10.1371/journal.pone.0037027; Amy X. Zhang, Anastasios Noulas, Salvatore Scellato, and Cecilia Mascolo, "Hoodsquare: Modeling and Recommending Neighborhoods in Location-Based Social Networks,"*arXiv* (2013); Justin Cranshaw, Raz Schwartz, Jason I. Hong, and Norman M. Sadeh, "The Livehoods Project: Utilizing Social Media to Understand the Dynamics of a City" (paper presented at the International Conference on Weblogs and Social Media, Dublin, Ireland, June 4, 2012).
9. *Livehoods*, accessed September 25, 2013, www.livehoods.org.

10. Rowan Wilken, "Mobilizing Place: Mobile Media, Peripatetics, and the Renegotiation of Urban Places," *Journal of Urban Technology* 15, no. 3 (2008): 39–55.
11. Lindsay T. Graham and Samuel D. Gosling, "Can the Ambiance of a Place Be Determined by the User Profiles of the People Who Visit It?"*Proceedings of the Fifth International AAAI Conference on Weblogs and Social Media* (Palo Alto, CA: AAAI Press, 2011).
12. Nadav Hochman and Lev Manovich, "Zooming into an Instagram City: Reading the Local Through Social Media,"*First Monday* 18, no. 7 (2013), accessed September 25, 2013, http://firstmonday.org/ojs/index.php/fm/article/view/4711/3698.
13. Keith N. Hampton, Oren Livio, and Lauren Sessions, "The Social Life of Wireless Urban Spaces: Internet Use, Social Networks, and the Public Realm," *Journal of Communication* 60, no. 4 (2010): 701–722.
14. Lee Humphreys and Tony Liao, "Foursquare and the Parochialization of Public Space," *Selected Papers of Internet Research* (2013) accessed November 24, 2013, http://spir.aoir.org/index.php/spir/article/view/784; Raz Schwartz, "Online Place Attachment: Exploring Technological Ties to Physical Places," *Location, Mobile Technologies and Mobility* (New York: Routledge, forthcoming).
15. Instagram, *Instagram Today: 150 Million People* (2012), accessed November 24, 2013, http://blog.instagram.com/post/60694542173/150-million.
16. Yelp, *About Yelp* (2013), accessed September 25, 2013, www.yelp-press.com/phoenix.zhtml?c=250809&p=irol-press.
17. Henri Lefebvre, *Rhythmanalysis: Space, Time and Everyday Life*, trans. Stuart Elden and Gerald Moore (London: Continuum, 2004).
18. Bryant Park, *Early History* (2013), accessed September 25, 2013, www.bryantpark.org/about-us/history.html.
19. Aaron Smith, "Smartphone Ownership 2013," *PEW Research Center* (June 5, 2013), accessed September 25, 2013, http://pewinternet.org/Reports/2013/Smartphone-Ownership-2013.aspx; Kathryn Zickuhr, "Location-Based Services," *PEW Research Center* (September 12, 2013), accessed September 25, 2013, www.pewinternet.org/Reports/2013/Location.aspx.
20. Ray Oldenburg, *The Great Good Place: Cafés, Coffee Shops, Bookstores, Bars, Hair Salons, and Other Hangouts at the Heart of a Community* (New York: Marlowe, 1999).
21. Larissa Hjorth, "Relocating the Mobile: A Case Study of Locative Media in Seoul, South Korea," *Convergence* 19, no. 2 (2013): 237.
22. Louise Barkhuus, Barry Brown, Marek Bell, Scott Sherwood, Malcolm Hall, and Matthew Chalmers, "From Awareness to Repartee: Sharing Location Within Social Groups" (paper presented at the Conference on Human Factors in Computing Systems, Florence, Italy, April 5–10, 2008).
23. Henriette Cramer, Mattias Rost, and Lars Erik Holmquist, "Performing a Check-in: Emerging Practices, Norms and 'Conflicts' in Location-Sharing Using Foursquare" (paper presented at MobileHCI 2011, August 30–September 2, 2011, Stockholm, Sweden).
24. Lee Humphreys, "Connecting, Coordinating, Cataloguing: Communicative Practices on Mobile Social Networks," *Journal of Broadcasting & Electronic Media* 56, no. 4 (2012): 494–510.
25. Lyn H. Lofland, *The Public Realm: Exploring the City's Quintessential Social Territory* (Hawthorne, NY: Aldine de Gruyter, 1998).
26. Bernie Hogan, "The Presentation of Self in the Age of Social Media: Distinguishing Performances and Exhibitions Online," *Bulletin of Science, Technology, and Society* 30, no. 6 (2010): 377–386.

5 Locative Praxis

Transborder Poetics and Activist Potentials of Experimental Locative Media

Andrea Zeffiro

> Every praxis has two historical co-ordinates: one denotes the past, that which has been accomplished, the other the future onto which praxis opens and which it will create.[1]

On September 1, 2010, Glenn Beck—on his former Fox News Channel program, *The Glenn Beck Show*—directed his viewers' attention toward the Transborder Immigrant Tool (TBT). This was a project to emerge from the University of California San Diego (UCSD), the aim of which was to deposit GPS-enabled disposable phones along the United States–Mexico border. These phones were equipped with a GPS application—a digital compass and an interactive map updated in real time—to assist migrants to orient themselves and locate resources, such as water caches and safe locations, while traveling within the border region. In the segment, Beck describes how "employees at the University of California San Diego are openly, and with the help of your hard earned dollars, aiding illegal aliens with the help of GPS cell phones." Beck warns: "America, this is madness and you know it. Common sense says we must turn the money off on this project and others like it. You can teach whatever you want but not with damn tax dollars . . . I think these people should be fired. But you can't. Tenure, you know."[2] By the end of the segment, the TBT and its researchers had been aligned with transnational terrorist networks and deemed a threat to the security, sovereignty, and identity of the United States.

Even prior to Beck's segment, three Republican congressmen in San Diego County had written letters to UCSD questioning both the aim and the scope of the project. In response, the university temporarily closed the b.a.n.g. lab,[3] the research collective directed by Ricardo Dominguez at the California Institute for Telecommunications and Information Technology, UCSD (Figure 5.1). Dominguez had created the TBT in collaboration with UCSD colleagues Brett Stalbaum, Micha Cárdenas, Jason Najarro, Amy Carroll, and Elle Merhmand. Additionally, the UCSD administration, the three Republican congressmen, and the FBI Office of Cybercrimes launched an investigation into Dominguez's work. Untangling the issues surrounding the investigation, Brett Stalbaum[4] has detailed how the administration

Figure 5.1 Ricardo Dominguez during a research trip to Calexico, California, a town on the U.S.–Mexican border, in 2008 (photograph by Brett Stalbaum).

threatened Dominguez's tenure because of a virtual sit-in he launched against the website of the University of California President's Office.[5] This virtual sit-in, which was held on March 4, 2010, in conjunction with a day of protest against fee increases at the University of California, was construed as a botnet,[6] even though Dominguez had held virtual sit-ins in the past without repercussions and had even been granted tenure for such work.[7]

It would be an understatement to describe the Transborder Immigrant Tool as highly contested. Indeed, the project has received a significant amount of attention throughout its development. Publicity aside, what is compelling about the work is the subtle and poetic manner in which the project confronts the fallacies of border spaces under global capitalism and connects to a larger trajectory of artistic and activist works intervening within and across border spaces and migrant narratives. The ideological concerns embedded within the Transborder Immigrant Tool—such as the contradiction between symbolic and physical borders and the processes of political subjugation and alienation—are preoccupations channeled through the work of Polish-born and American-based media artist and designer Krzysztof Wodiczko. Known for his projections, instruments, and vehicles, a number of Wodiczko's works confront im/migrant experiences. *Alien Staff* (1992–1993),[8] for instance, is a communicative intervention into the "preconceived categories of strangeness and difference" that frame and silence immigrants.[9] The staff, which resembles "a biblical shepherd's rod," is equipped with a mini monitor and speaker and "broadcasts" narratives and images that speak to the immigrant experience.[10] On the one hand, "the presence of the staff-carrying alien in real space and time makes clear and denounces the inability of media images to depict the experience of the exile and his or her memory."[11] On the other hand, the instrument dealienates and instigates communication: as it is carried through an urban space, the individual immigrant is afforded an opportunity to "address" directly anyone intrigued by it.[12] *Alien Staff* serves as a surrogate public space, and Wodiczko describes the work as "a form of portable public address equipment and cultural network for individuals and groups of immigrants."[13] As an interventionist tool, *Alien Staff* redirects public attention to the issues and the groups of individuals who are otherwise misrepresented, underserved, or rendered invisible in public spaces.

In many ways—both thematically and technologically—Wodiczko's project is a precursor to the TBT and to what Rowan Wilken describes as the "more experimental, speculative and emerging areas of mobile media development."[14] In what follows, the Transborder Immigrant Tool is employed as a case study toward the refinement of locative praxis: a conceptual framework for understanding the ways in which experimental locative media might engage in political and cultural activism and dissent. As a conceptual framework, locative praxis articulates a politicized dimension of locative media through which the dialectic of practice and reflection, at the intersection of social action and intent, can further an understanding of the spatial and sociopolitical dimensions of a particular space/place.

LOCATIVE MEDIA

"Locative media" is a descriptive term that designates the use of an assemblage of mobile and location-aware technologies—notably GPS and cellular telephony—in the production of site-specific experiences, performance pieces, interactive art works, and public installations. The term has been used to describe location-based services and commercial applications, which span fleet tracking, in-car navigation, and emergency services, to geotagging content through social networking platforms like Facebook and Twitter. Central to the facilitation of both commercial services and applications and of locative artistic and cultural works is the Global Positioning System (GPS): a worldwide satellite navigational system that was developed by the United States Department of Defense and is maintained by the United States Air Force. Over the course of the last ten years, consumer-grade GPS has transitioned from an external specialized device to an embedded (that is, out-of-site) component within mobile digital technologies, perhaps most strikingly with the seamless integration of GPS into mobile phones.

Yet what sets the Transborder Immigrant Tool apart from location-based services and commercial applications is an engagement with and an articulation to Lisa Parks' query, "how might Western controlled satellite technologies be appropriated and used in the interests of a wider range of social formations?"[15] Parks' question also intimates an underlying tension of locative media, namely that these artistic and cultural works not only appropriate but also assimilate military and commercial technologies and applications. Locative praxis is advanced here as a conceptual framework to tease apart this friction, confront the political issues and relationships organizing the spaces of locative media discourse, and cultivate a continuous exchange between locative media and activist potentialities.

LOCATIVE PRAXIS

Although "praxis" shares its etymological links with "practice," it does not carry the same semantic undertones.[16] For Henri Lefebvre, praxis "comprises a theoretical decision as well as the decision to act"[17] and "always points to the domain of possibility."[18] In this regard, praxis is more than an interpretation of the world. It not only demands reflection upon and action within the world in order to transform it, it also entails a critical inflection toward a future that is marked by possibility. Locative praxis, therefore, is a critical and interpretive framework for analyzing the ways in which experimental locative media might intervene in the spatial politics of a space/place.

Indeed, spatial metaphors continue to organize both discourses and practices associated with locative media. The term itself—"locative media"—is an appropriation of the locative noun case from the Latvian language, which indicates a final location of action or a time of the action, and it loosely corresponds to the English prepositions "in," "on," "at," "by."[19] Under the rubric of locative praxis, however, location represents all that has become implanted in the social production of that particular space, as well as the tensions accompanying the relations of power embedded within the technological devices and infrastructures that facilitate these works.

Location, much like praxis, is a productive process.[20] Lefebvre approaches the theorization of the formation of social space vis-à-vis a spatial triad that is marked by spatial practice (perceived space), the representation of space (conceived space), and representational space (lived space).[21] The triad intimates the manner in which any (social) space is reflective of the dominant modes of production of its historical condition. "The shift from one mode [of production] to another," explains Lefebvre, "must entail the production of a new space."[22] The development of modern Western capitalism parallels the production of what Lefebvre refers to as "abstract space." The aim of Lefebvre's project, therefore, is the establishment of a "differential space," a space that subdues the forces of homogenization within abstract space—or spaces of capitalism. Within this purview, the Transborder Immigrant Tool facilitates the production of revisionist border narratives that simultaneously accentuate the contradictory political and economic effects of capitalist globalization[23] and that challenge the conflation of border control with the autonomy, sovereignty, and safety of a nation.

FRAME OF REFERENCE

When the United States, Canada, and Mexico signed the North American Free Trade Agreement (NAFTA) in 1994, supporters looked favorably on the elimination of barriers to trade and investment as a course toward economic growth. Mexico's potential to thrive economically under NAFTA, however, translated differently in practice. Many industries—particularly textile and garment—moved south, where labor was cheaper. At the same time, citizens from rural and indigenous regions of Mexico migrated north—to border regions[24]—to find work in factories. Over one million Mexicans have worked in these factories and plants, or *maquiladoras*. Along with exploitive and unsafe working conditions, low wages, and job insecurity, workers routinely encounter violence, sustain injuries from unsafe machinery and toxic chemical exposure, and are barred from any kind of labor organizing.[25] As the continental divisions blurred for trade, border control became increasingly more mechanized. As a contested site, explains Ian Alan Paul, "[t]he border between the U.S. and Mexico functions simultaneously to ensure the steady uninterrupted flow of capital and goods while also denying unpermitted or undocumented crossings."[26] As a consequence of the mobility of capital and the transnationalization of production processes, labor has become increasingly mobile, especially with the outmigration of labor from countries of the global South. Nevertheless, global migrant labor is overwhelmingly perceived as a threat to the safety and sovereignty of a nation.[27]

When Arizona Governor Jan Brewer signed the controversial Support Our Law Enforcement and Safe Neighborhoods Act (Senate Bill 1070)[28] in 2010, high levels of illegal immigration and crimes committed by unauthorized immigrants were among the key rationales cited. Proponents of the bill, however, overlooked two salient points. According to a report published by the nonpartisan Immigration Policy Institute, "[c]rime rates have already been falling in Arizona for years despite the presence of unauthorized immigrants, and a century's worth of research has demonstrated that immigrants are less likely to commit crimes or be behind bars than the native-born."[29] And while the establishment of post-NAFTA security measures, like Operation Gatekeeper in California, Operation Hold-the-Line in Texas, and Operation Safeguard in Arizona, has contributed to a concentrated effort on the border itself, other regions, especially the harsh zones in the borderlands/*la frontera*, remain unsecured. Nearly 3,000 migrants have died[30] from exposure in these unsupervised zones while attempting to traverse the U.S.–Mexican border.

The use by migrants of mobile technologies—including encrypted radios and cell phones—within border regions is common. Migrant traffickers, known as coyotes, hide out in mountainous terrain and remotely instruct migrants crossing into the United States illegally. Instructions are relayed via cell phone, and the directives pertain to ways of evading the Border Patrol.[31] For such a service, undocumented migrants pay thousands of dollars, with the cost depending on the distance of the smuggling route and "add-ons," such as a guaranteed passage back into the United States for the apprehended or fake or stolen visas and passports. The transportation or smuggling of so-called illegals has become a lucrative business venture that has spawned well organized syndicates.[32] Yet despite the coordinated efforts, there is little if any guarantee that migrants receive what they pay for or survive the journey.

BORDER POETICS

It is at this juncture—between real and symbolic forms of violence enacted on migrants—where the Transborder Immigrant Tool intervenes (Figure 5.2). For migrants crossing border spaces, the tool provides support in the form of a navigational aid that supplies directional cues within what is otherwise

Figure 5.2 The Transborder Immigrant Tool billboard for Galeria de la Raza in San Francisco (photograph by Ricardo Dominguez).

a barren and harsh landscape. It therefore offers a virtual shelter: it communicates directions toward water caches and safe locations.[33] The TBT can also be perceived, both theoretically and practically, as border poetics, that is, as the cultural expression or figuration of border-crossing narratives. To read the Transborder Immigrant Tool in this way, as a figuration of border poetics, is to perceive the project as a cultural expression of border formations and experiences.

Describing border poetics as "sensitive to borders on many levels,"[34] Johan Schimanski applies border poetics to the study of border-crossing narratives, the five most important being, as the author proposes:

> *[T]extual borders*, the edges and divides of the text as discourse; *topographical borders*, borders in the represented space of the setting; *symbolic borders* or, properly, differences which have allowed themselves to be territorialized in a symbolic landscape; *temporal borders*, the passages from one period to another—territorialized time; and *epistemological borders*, borders which positivism imagines as always expanding—knowledge territorialized.[35]

These categories support a reading of border narratives as multitudinous and diffuse, in that these spatial (or border) planes open onto two historical coordinates: one that denotes the past and that has been accomplished and the other onto the future and toward the creation of a differential border space.[36]

To fully comprehend the border poetics expended by the TBT, it is necessary to acknowledge its social backdrop and historical coordinates. As mentioned, the Tool is a project developed by members of the b.a.n.g. lab at the California Institute for Telecommunications and Information Technology at UCSD. However, prior to the institution of the lab or the research on and development of the TBT, Dominguez and Stalbaum, in addition to Stefan Wray and Carmin Karasic, cofounded the Electronic Disturbance Theatre (EDT)[37] in 1997. The EDT was responsible for the creation of digital artifacts and acts of nonviolent protest or "electronic civil disobedience,"[38] and their efforts concentrated on interceding in colonized spaces, notably cyberspace.

One of the earliest and perhaps most iconic acts of electronic civil disobedience orchestrated by the EDT was a virtual sit-in organized in solidarity with the Zapatista movement in Chiapas, Mexico. In 1998, the EDT released the Zapatista FloodNet, an iteration of the Java applet employed in the virtual sit-in launched against the website of the University of California President's Office in 2010. Created to interrupt the website of the Mexican government as a means of protesting the oppression of the indigenous people in Chiapas, the EDT subsequently released FloodNet as open source software. During "the heyday of the anti-globalization movement," explains Marco Deseriis, "hundreds of thousands of hacktivists launched it through their browsers to block or slow down access to the WTO, IMF, World Bank, and World Economic Forum websites."[39] Thus, the TBT is situated along an axis of software tool development for acts of electronic civil disobedience.

The software at the center of the TBT is Brett Stalbaum's program, the Virtual Hiker, an algorithm that translates geospatial terrain into a virtual trail, or hike. Using GPS satellite signals, the algorithm identifies routes within a block of map data and creates a so-called bread crumb trail showing the locations and distances traveled on a map. Intrigued by the capabilities of the Virtual Hiker, the collective explored the manner in which the program could be hacked to assist migrants crossing border zones. "We asked ourselves," recounts Dominguez, "what were the spaces of necessity or danger on the border, and how could we plug in this new element of the GPS structured cell phone?"[40] When combined with a GPS-enabled cell phone, the Virtual Hiker program plots one's place at any particular location on a map and traces—in real time—a path as movement occurs.

In 2005, the EDT 2.0 released the first iteration of the Tool, which was initially prototyped on the Motorola i455 phone due to its economical cost—roughly US$40—and because its GPS functionality worked independently of a service plan. Circumventing a service plan is important for two reasons. First, it lowers the total cost of the Tool; second, and perhaps more importantly, it bypasses the cellular or mobile network. When on a network, a phone registers its position with cell towers every few minutes, whether the phone is being used or not, and mobile carriers can and often do retain data on their customers.[41] While sidestepping a service plan means that calls cannot be placed from the phone, the evasion of a phone carrier adds a layer of protection. So long as the GPS function of the phone remains on, the device can maintain constant communication with satellites and thus receive directional cues via the Virtual Hiker program (Figure 5.3).

As for the Tool's interface, the collective chose a navigational aid that resembles a traditional compass, and they employed the vibration function of the phone as an alert mechanism. The interface design speaks to the fact that many migrants are from indigenous communities and speak neither Spanish nor English. The compass therefore offers directional cues without the burden of translating language, while the vibration function enables migrants to focus on the actual terrain as opposed to checking the phone for updates. Although the tool is not yet widely available—the original device proved to be unreliable, and the team continues to test for inexpensive mobile phones with hackable GPS to support the Virtual Hiker—the aim is to refine the tool into a distributable prototype that can be circulated by migrant aid organizations. A website (walkingtools.net) was launched to encourage take-up of the tool. The site catalogs the development of the open-source code and allows for the appropriation of the toolkit for use in other border regions around the world.

Figure 5.3 The Transborder Immigrant Tool in operation—with a screenshot from a Nokia e71 cell phone—directing a user to a Water Station Inc. water cache in the Anza Borrego Desert, located within the Colorado Desert of southern California (photograph by Brett Stalbaum).

Describing both the functionality and symbolism of the Transborder Immigrant Tool, Ricardo Dominguez explains that "[i]t is a tool that not only allows a sort of safety but also creates this kind of deeply poetic centering of hospitality. In a certain sense, we think of it as almost the Statue of Liberty that one carries around as one walks."[42] A description of the tool as "deeply poetic" suggests that, in spite of its utilitarian function, the TBT has imaginative and emotional potentialities and is therefore engrained within a symbolic narrative, in which the notion of hospitality corresponds to the emblem of international friendship epitomized by the Statue of Liberty. Viewed through the purview of political agency, specifically as an artivist[43] statement on the part of its developers, the tool, as an invention to assist migrants cross border regions, is an act of compassion, kindness, and, indeed, hospitality.

A symbolic alignment between the TBT with the Statue of Liberty also illuminates the contradictions of immigration policies and border control under global capitalism. Frederic Auguste Bartholdi's statue, *Liberty Enlightening the World*, represents Libertas, the Roman goddess of freedom. At her feet lie the broken shackles of tyranny and oppression, and in her arm she bears a torch and a tabula ansata inscribed with the date of the American Declaration of Independence. The statue's proximity to Ellis Island—a small island in the port of New York and New Jersey that served as an immigrant inspection station from 1892 to 1954—positioned the statue as a welcoming beacon for the millions of immigrants arriving in the

United States. And the statue continues to serve as an icon of freedom and democracy, despite the fact that neither freedom nor democracy is equally dispersed, and the authorization to cross a border space depends not only on a passport but on the right kind of passport.[44]

The continuity and cohesion of a symbolic border space, especially one's perception of and subsequent subjectification by it—as legal citizen or illegal alien—is dependent on what Peter Andreas describes as "escalating symbolic performance." Within this purview, border control functions as a kind of political stage for public performance.[45] "'Successful' border management," explains Andreas, "depends on successful image management [that] does not necessarily correspond with levels of actual deterrence."[46] As a result, border control efforts "are not only *actions* (a means to a stated instrumental end) but also *gestures* that communicate meaning."[47] In this regard, the border (space) exerts and maintains control as long as representations of (border) space continue to generate what Wendy Brown describes as "performative and symbolic effects in excess of their obdurately material ones."[48] In other words, as long as the idea of the "border" supersedes knowledge of the real consequences or inadequacies of border management, then the "border"—as a thing, or concept—will continue to serve a purpose and have resonance within the collective unconscious.

The historical framing of the Transborder Immigrant Tool situates the project within larger border-crossing narratives, notably temporal borders, epistemological borders, and symbolic borders. In terms of the temporal, the project is a coordinate along an axis of software tool development for acts of electronic civil disobedience. As for epistemology, the TBT is materialized through the transfer of knowledge within an active network of hacktivists and artivists. And, finally, in relation to the symbolic, the TBT draws out the contradictions between symbols of immigration and the lived realities of so-called illegals. For certain bodies, border spaces exist as passageways that separate one territory from another. For other(ed) bodies, however, the border is a site of death, or detention, or a passageway where one's unauthorized or illegal permeation signifies a life of anonymity or secrecy. Reading through the lens of border poetics, the TBT communicates narratives of temporality, epistemology, and symbolism, which anchor the project to its first historical coordinate—the past, or to that which has been accomplished.

BORDER SPACES

As an intervention within border politics and across multiple performative and symbolic spaces, the TBT circulates within what Peter Andreas describes as the "escalating symbolic performance" of border control (Figure 5.4). Extending Andreas's argument, Rita Raley maintains that artistic interventionist tactics that take aim at border control must be understood

Figure 5.4 Brett Stalbaum shortly after he and Ricardo Dominguez had been pulled over by the border patrol on Highway 98 near Calexico; Stalbaum had just finished talking to the officers (photograph by Ricardo Dominguez).

in the same terms, as "a battle that is at once material and symbolic, fought on the very 'political stage' where power is exercised."[49] In other words, artistic interventions are not merely a critical response to the manifestation of power but also a material consequence of it.

The production of a space that takes aim at border spaces is therefore indebted to the reproduction of those spatial (border) practices. Continuity and cohesion of dominant representations of border spaces are necessary if the TBT is to be read against dominant narratives. For Lefebvre, representations of space are the official spaces within any society, and they correspond to the modes of production, as well as to "the 'order' which those relations impose, and hence to knowledge, to signs, to codes, and to 'frontal' relations."[50] If representations of border spaces correspond to their mode of production—global capitalism—and the order that those relations impose—a conflation of border control with the autonomy, sovereignty, and safety of a nation—then the Transborder Immigrant Tool renders visible both the modes of production and the order that those relations impose.

A reordering of these relations occurs in the space of "inhabitants" and "users," that is, in lived space, or what Lefebvre defines as "space as directly lived through its associated images and symbols."[51] This is also the space of lived experience, which emerges from the dialectical relationship between spatial practice and representations of space. Therefore, it is at this point of tension, between perceptions and conceptions of the border space on the one hand and the counternarratives provided by the TBT on the other, wherein a differential space emerges. Under contemporary global capitalism, border spaces, like the abstract spaces of capitalism referenced by Lefebvre, carry the seeds or potentialities for a new kind of space—a differential border space. The differential border space to emerge alongside the Transborder Immigrant Tool challenges the conflation of border control with the autonomy, sovereignty, and safety of a nation and reveals the fallacy of border spaces under global capitalism. The production of a differential border space anchors the TBT to its second historical coordinate: the future onto which praxis opens and that it will create.[52]

A recent experimental locative media work, called *Border Bumping*, links the Transborder Immigrant Tool to this second historical coordinate. Described as "dislocative" media, Julian Oliver's project, *Border Bumping*,[53] implements cellular telecommunication infrastructures as a "disruptive force"[54] to challenge the integrity of national borders. "When you are crossing a border between two countries, your mobile phone often connects to cell towers of the entering country before or after the actual political border, which results in seemingly distorted border lines," explains project collaborator Till Nagel. "This project records and visualizes these virtual bumps to get an impression of the new digital boundaries of the world."[55] In the way that Krzysztof Wodiczko's *Alien Staff* constitutes an antecedent to the TBT, *Border Bumping* represents a future locative praxis, which the TBT connects to as part of a larger trajectory of experimental locative media intervention and activism.

CONCLUSION

As a case study employed for the refinement of locative praxis, the Transborder Immigrant Tool exemplifies how mobile and location-based technologies, at the intersection of social action and intent, can further an understanding of the spatial and sociopolitical dimensions of border spaces. The TBT is first and foremost entrenched in the border politics confronting the United States and Mexico and therefore privileges the specific geographical locations and historical circumstances that are explicit to the relationship between those two countries. The issues embedded in the TBT, however, resonate on a global scale. The Tool facilitates a counternarrative to dominant border paradigms, and accentuates the contentious and consequential issues invested in the maintenance of border spaces under capitalist globalization.

Locative praxis, however, is not only a critical and interpretive framework for analyzing the ways in which experimental locative media might intervene within the sociopolitical issues of a particular space/place; it is also a mode of inquiry into the technological devices and infrastructures that facilitate a work. In this sense, the Transborder Immigrant Tool demonstrates how mobile technologies might be appropriated within activist formations, as tools for not only interaction but also reaction, that is, as instruments to facilitate political and cultural activism and dissent. Indeed, these technological devices and infrastructures are linked to military and commercial ventures. Yet the appropriation of mobile and location-based technologies, within initiatives invested in political and cultural activism, expands their range of use and adoption. In other words, experimental locative media works generate novel ways of perceiving the utility of these technologies, not only in terms of how these devices might be used but also, and perhaps more importantly, by whom.

NOTES

1. Henri Lefebvre, *The Sociology of Marx* (New York: Columbia University Press, 1982), 55.
2. Glenn Beck, *The Glenn Beck Program* (Fox News Channel, 2010).
3. Bits, atoms, neurons, and genes lab.
4. Brett Stalbaum, "Attack on Academic Freedom: Is the Persecution of Dominguez Really About the TBTool?" *Walking Tools*, accessed April 10, 2010, www.walkingtools.net/?p=437.
5. Regents of the University of California, *University of California: Office of the President* (2012), accessed September 13, 2013, http://www.ucop.edu/.
6. A bot is considered to be malicious software because it causes a computer to perform automated tasks over the Internet without the knowledge of the user.
7. For more on the controversy surrounding the threats to the lab and Dominguez's tenure, see Zach Blas, "Interview with Ricardo Dominguez," *Reclamations Blog* (January 17, 2012), accessed September 13, 2013, www.reclamationsjournal.org/blog/?ha_exhibit=interview-with-ricardo-dominguez; Leila Nadir, "Poetry, Immigration and the FBI: The Transborder Immigrant Tool," *Hyperallergic* (July 23, 2012), accessed September 13, 2013, http://hyperallergic.com/54678/poetry-immigration-and-the-fbi-the-transborder-immigrant-tool/.
8. Other works by Wodiczko include *Mouthpiece* (1993–1997) and *Tijuana Projection* (2001).
9. Krzysztof Wodiczko, *Alien Staff*, accessed September 13, 2013, http://stuff.mit.edu/afs/athena/course/4/4.395/1995_retired/www/krystof/krystof.html.
10. Wodiczko, *Alien Staff*.
11. Marie Fraser, "Media Image, Public Space, and the Body: Around Krzysztof Wodiczko's *Alien Staff*," in *Precarious Visualities: New Perspectives on Identification in Contemporary Art and Visual Culture*, edited by Olivier Asselin, Johanne Lamoureux, and Christine Ross, trans. Timothy Barnard (Montréal: McGill-Queen's University Press, 2008), 256.
12. Wodiczko, *Alien Staff*.
13. Wodiczko, *Alien Staff*.
14. Rowan Wilken, "A Community of Strangers? Mobile Media, Art, Tactility and Urban Encounters with the Other," *Mobilities* 5, no. 4 (2010): 450.
15. Lisa Parks, *Cultures in Orbit: Satellites and the Televisual* (Durham, NC: Duke University Press, 2005), 10.

16. Adolfo Sánchez Vázquez, *The Philosophy of Praxis*, trans. Mike Gonzalez (Atlantic Highlands, NJ: Humanities Press, 1977), 1.
17. Lefebvre, *The Sociology of Marx*, 54.
18. Lefebvre, *The Sociology of Marx*, 55.
19. Karlis Kalnins, "[Locative] Locative Is a Case Not a Place" *[Locative] Listserv* (May 10, 2004), accessed September 13, 2013, http://web.archive.org/web/20050907080544/http://db.x-i.net/locative/2004/000385.html.
20. "Productive" here implies an investment in action or doing with the aim of transformation.
21. Henri Lefebvre, *The Production of Space*, trans. Donald Nicholson-Smith (Oxford: Blackwell Publishers, 1991).
22. Lefebvre, *The Production of Space*, 46.
23. Neferti Tadiar, "Introduction: Borders on Belonging," *The Scholar & Feminist Online* 6 no. 3 (2008), accessed September 13, 2013, http://sfonline.barnard.edu/immigration/tadiar_01.htm.
24. United States–Mexico border states: Texas, New Mexico, Arizona, California, Baja California, Sonora, Chihuahua, Coahuila, Nuevo León, and Tamaulipas.
25. David Bacon, *The Children of NAFTA: Labor Wars on the U.S./Mexican Border* (Berkeley: University of California Press, 2004).
26. Ian Alan Paul, "Border Politics, Border Poetics" (MA diss., San Francisco Art Institute, 2011), 12.
27. Neferti Tadiar, "Introduction: Borders on Belonging."
28. The bill was the strictest anti-immigration law to be passed in the United States and expressed an "attrition through enforcement" doctrine. The legislation stipulated that it was a state misdemeanor for immigrants not to carry proper documentation at all times when in Arizona, as were hiring, sheltering, or transporting unregistered individuals. Under the bill, Arizona law enforcement officials were required to determine an individual's immigration status during a lawful stop or contact or when there is reasonable suspicion that an individual is an illegal immigrant. And state and local officials or agencies were barred from restricting enforcement of federal immigration laws. The day before the bill was to become law, a federal judge issued an injunction that blocked numerous provisions outlined in the bill. On July 25, 2012, the U.S. Supreme Court issued a ruling in *Arizona v. United States,* which upheld the portion of the bill authorizing Arizona law enforcement officials to investigate the immigration status of individuals during lawful contact.
29. Cited in Brad Warbiany, "Arizona Crime Stats," *IndyTruth.org* (May 3, 2010), accessed September 13, 2013, www.indytruth.org/archive/2010/0503-tlp-immigration.html.
30. Rita Raley, *Tactical Media* (Minneapolis: University of Minnesota Press, 2009), 32.
31. "Traffickers, Migrants Use Cell Phones Along U.S.–Mexico Border," *Latin American Herald Tribune* (September 13, 2013), www.laht.com/article.asp?CategoryId=14091&ArticleId=446403.
32. Martin Brass, "The U.S./Mexican Border Has Become a Sieve of Death," *Military.com* (2004), accessed September 14, 2013, www.military.com/NewContent/0,13190,SOF_0804_Mexico,00.html.
33. The mapping of water caches and safe zones was completed with the assistance from migrant aid NGOs Border Angels and Water Station Inc. See http://borderangels.org and http://www.desertwater.org/.
34. Johan Schimanski, "The Postcolonial Border: Bessie Head's 'The Wind and a Boy,'" in *Readings of the Particular: The Postcolonial in the Postnational*, eds. Anne Holden Ronning, and Lene Johannessen (Amsterdam: Editions Rodopi B.V., 2007), 73.

35. Schimanski, "The Postcolonial Border," 73.
36. See Lefebvre, *The Sociology of Marx*: 55.
37. The current iteration of the EDT is known as the EDT 2.0.
38. Stefan Wray, "On Electronic Civil Disobedience" (paper presented at the Socialist Scholars Conference, New York, March 21–22, 1998), www.thing.net/~rdom/ecd/oecd.html).
39. Marco Deseriis, "Online Activism as a Participatory Form of Storytelling," in *Art and Activism in the Age of Globalization*, eds. Lieven De Cauter, Ruben De Roo, and Karel Vanhaesebrouck (Rotterdam: NAi Publishers, 2011), 255.
40. Dominguez quoted in Corinne Ramey, "Artivists and Mobile Phones: The Transborder Immigrant Project," *MobileActive.org* (November 18, 2007), http://mobileactive.org/artivists-and-mobile-pho.
41. American Civil Liberties Union, "Cell Phone Tracking Request Response—Cell Phone Company Data Retention Chart," ACLU (September 27, 2011), www.aclu.org/cell-phone-location-tracking-request-response-cell-phone-company-data-retention-chart.
42. Beck, *The Glenn Beck Program*.
43. Guisela Latorre, "Border Consciousness and Artivist Aesthetics: Richard Lou's Performance and Multimedia Artwork," *American Studies Journal* 57 (2012), www.asjournal.org/archive/57/216.html.
44. The Obama administration has deported over 1 million people, more than any president since the 1950s.
45. Peter Andreas, *Border Games: Policing the U.S.–Mexico Divide* (Ithaca, NY: Cornell University Press, 2001), 9.
46. Andreas, *Border Games*, 11.
47. Andreas, *Border Games*, 11.
48. Wendy Brown, *Walled States, Waning Sovereignty* (Brooklyn, NY: Zone Books, 2010), 39–40.
49. Raley, *Tactical Media*, 34.
50. Lefebvre, *The Production of Space*, 33.
51. Lefebvre, *The Production of Space*, 39.
52. Lefebvre, *The Sociology of Marx*, 55.
53. Julian Oliver, "Introduction," *Border Bumping*, http://borderbumping.net/.
54. Till Nagel, "Border Bumping," *TillNigel.com* (September 21, 2012), accessed September 13, 2013, http://tillnagel.com/2012/09/border-bumping/.
55. Nagel, "Border Bumping."

Part 2

Geography, Code, Representation

6 Map Interfaces and the Production of Locative Media Space

Jason Farman

INTRODUCTION

When Apple launched the iOS 6 version of their operating system for the iPhone to correspond with the launch of the iPhone 5 in September 2012, the feature that brought the most attention was the removal of Google Maps, which was replaced by Apple's new in-house Maps app. Initially, Google Maps was not only removed from the system but was also unavailable for downloading by users who preferred it to Apple Maps. Within hours of the launch of iOS 6, the firestorm of complaints began to spread across the Internet. Within 48 hours, users had posted 245 screen captures from their iPhones to the Tumblr page, *The Amazing iOS 6 Maps*, showing absurd glitches and misdirections from the Apple Maps app. These posts included a wide range of errors from geographical distortions in the Apple Maps 3D perspective to driving directions that took people across oceans and airport runways. Apple was also criticized for not including directions for users of public transportation, a move commented on by the @iOS-6Maps satire Twitter account: "Look, we left out transit directions because only losers take public transport. All iPhone owners drive BMWs, everyone knows that."[1] Most critiques that echoed throughout the web in the days and months that followed the release of Apple Maps resonated with Anil Dash's argument in his article, "Who Benefits from iOS6's Crappy Maps?" in which he wrote, "Apple made this maps change despite its shortcomings because they put their own priorities for corporate strategy ahead of user experience."[2]

Although frustration over the prioritization of corporate interests over user experience is an important perspective for us to understand the public outrage over Apple Maps, it is not, in my estimation, at the center of what caused the widespread discontentment with the new map interface. Instead, the key to the uproar over the shift in mapping interfaces was that it was a change in default maps happening on *mobile devices*. Although such a transition would have undoubtedly brought criticism if it were a change happening on a MacBook, it would have not caused the same sort of outrage that the shift on a mobile operating system caused. This outrage over the "incorrectness" of Apple Maps (from sending people to the wrong city to showing the satellite view of their towns as nothing but large pixels or clouds) was brought about because mobile users are currently using maps as their default interface for navigating everyday space. This was not simply a change in one app; it was a change in the entire experience of using an iPhone as a means to orient oneself in space and to understand that space in a context-specific way.

The series of events that unfolded after the launch of iOS 6 and Apple Maps caused me to investigate the increasing role that map interfaces play in our practices of locative media and location-based services. Thus, of key concern for this chapter are two questions. Why are maps our default interface for many locative media projects? How do maps function as a tool to orient ourselves spatially when using locative media? In the context of these two inquiries, I am interested in our cultural orientation toward objects like maps and what ultimately is lost in the map-as-interface model of contemporary locative media.

THE OVEREMPHASIS ON SPATIAL DYNAMICS

To begin with an investigation into the relationship between orientation and mapping, I must first take a step back to consider the emphasis on the spatial in contemporary mobile computing culture. From Sarah Sharma's perspective, the focus—indeed, the overemphasis—on spatial dynamics in locative and mobile media not only relies on the most simplistic analysis of the media but also "risks aligning the field too closely with the logic of the market."[3] She argues, "That a media-technology alters the qualitative experience of time and space is hardly surprising [. . .] All forms of media-technology, whether they are mobile, locative, clunky, minuscule, or fixed in place will alter, regulate, and change the qualitative experiences of space and time."[4] Drawing on the work of Harold Innis, she goes on to argue that such an overemphasis on the spatial dynamics of mobile and locative media plays into the centralizing characteristics of imperial powers: "Civilizations that emphasize space over time tend to be imperial powers, involved in the conquering of space at the expense of the maintenance of culture over time . . . by all such determinations, global capital depends on spatially biased cultures. It is not just that our dominating technologies are spatially biased but our ways of knowing, systems of power, and even notions of resistance, tend to be spatial."[5] To address these concerns, it is worth asking, "Do the locative media projects discussed in this book emphasize spatial or temporal engagement?" While the answer to that question must indeed be "both," most discussions of locative and mobile media (like mine in this chapter) tend to emphasize the spatial dynamics of these media. Space and time, however, cannot be separated: space implies time (i.e., time is required to traverse a space, and duration is spatial in the ways that it is social/relational). Treating space and time as a binary instead of positioning them within a mutually dependent dialogical relationship tends to oversimplify our analyses of locative media.

Ultimately, Sharma's critique does not go so far as to argue for an emphasis on time over space; instead, she seeks to push scholars of space and place to interrogate *whose* space we are talking about. Because space, as Henri Lefebvre has famously noted, is social (and cannot be considered outside of the social), we must ask whose space is considered when looking at how locative media projects impact space (and how the space impacts these locative interfaces).[6] Sharma correctly asks, "Whose routes and paths do we valorize when we explain how mobile phones change space when we can play games on the bus or find a new pizza joint? Whose labor and time is re-orchestrated to make any of the 'new' things happen? What routes and paths are devalued and what regimes of dependency are created in these new techno-cultural practices?"[7]

Analyzing map interfaces in locative projects offers an important approach for thinking through this critique. In terms of the colonization of a culture (which, it can be argued, must be both spatial and temporal), an important structuring tool for such projects is the map. Along with such critiques, it is vital to inquire about the role of representations in our practices of space. Are we dependent on representational objects like maps when producing space? How do maps relate to our sense of embodiment when experiencing locative media projects?

EMBODIED MAPS

Maps are signifying tools for how we think about the world. They are also representations of the ways that we *want* to practice the world. Following Lefebvre's argument that space should not be considered as an empty container that we fill with our bodies (and instead should be understood as produced co-constitutively with bodies), the embodied practices of space are how it takes shape and is given meaning.[8] Space is produced and practiced; it is not a given (i.e., it does not exist *a priori*). Space always must be understood as existing within the cultural constraints that impact the ways that we practice space and even think about it.

As I argued in *Mobile Interface Theory*, the ways that we represent space are indelibly linked to how we embody and practice that space.[9] Representations not only frame our thinking about a space but also serve as modes of spatial production that make the space fit with our understanding of the world. There is thus a dialogical relationship between representational objects like maps and our embodied engagement with spaces. It is therefore not surprising that many locative media projects use maps as their primary interface. What *is* surprising is the use of maps in such body-centric media like locative media despite the fact that maps have traditionally offered representations of space as "objective." That is, maps (if they are "correct") are understood as standing in as an objective, often scientifically produced, index of reality (to borrow a phrase from Charles Sanders Peirce). Peirce's semiotic approach argued that "representations have power to cause real facts,"[10] and maps (especially those comprised of GPS or aerial photography) are indexes of reality, "in contrast to the icon's relatively straightforward resemblance and the symbol's conventionality or arbitrariness."[11]

An example of a locative media art project can help concretize these tensions between the map as a representation of objective space and the emphatic embodiment of locative media. In 2010, the Museum of London released a locative app called Streetmuseum, which took a large number of the museum's photographs and paintings about London and turned them into site-specific interactions. Each photograph or painting corresponds with the coordinates of the image's point of view. For example, someone using the app could walk to the South Bank of the Thames just west of Waterloo Bridge and bring up a painting from 1684 titled, *A Frost Fair on the Thames at Temple Street*. When the user clicks on this placemarker on the map, the painting loads onto the screen and can be overlaid onto the current perspective of the Thames. Using this augmented reality (AR) approach, in which the painting is somewhat transparent and can be held up to a current view of the scene captured by the phone's camera, this historic image of a fair on the Thames—when it was frozen over and "entertainment booths, coaches, sledges, sedan chairs and people can be seen on the ice," as the caption says in the app—directly corresponds (and often contradicts or stands in stark juxtaposition) to the real-time view of the place. A memorable scene for me was standing near 23 Queen Victoria Street and bringing up a photograph of the Salvation Army International Headquarters crashing to the ground soon after a night raid during the Blitz of World War II. The app thus seeks to take the Museum's collections to the streets, offering users a site-specific way of interacting with London's history.

The app serves as a good example of the tension between the map as an objective perspective of the world and locative media as intensely embodied due to its prioritizing of the individual's perspective. When encountering the app for the first time, you are presented with a top-down map of London with place markers scattered about, signifying locations where a photograph can be viewed. Users alternate between this top-down perspective (what Donna Haraway understood as a master–subject perspective of "seeing everything from nowhere"[12]) and the photograph/painting taken from a particular embodied perspective. This tension between the objective representation of space and the embodied perspectives on a space is an important entry point into an analysis of the use of maps in locative media.

While historical maps did offer a sense of representing space objectively (i.e., trying to capture, through scientific approaches, the accuracy of the landscape), contemporary maps have increased their perceived objectivity due to the data-gathering methods and modes of photographic representation. Because many of our maps are represented through satellite visualization, a mode of capture that is perceived to be detached from human agency, inasmuch as the imagery is taken not from a human's point of view but from a machine's, it can be (and has been) assumed that maps are an index of reality.[13] Yet theorists and cultural geographers have sought to trouble this alignment by extending or augmenting Jean Baudrillard's famous aphorism, "The territory no longer precedes the map, nor does it survive it."[14] As Rob Kitchin, Martin Dodge, and Chris Perkins have noted, many cultural geographers have followed Baudrillard by arguing that "a territory does not precede a map, but that space becomes territory through bounding practices that include mapping." They continue, "And since places are planned and built on the basis of maps, so space is itself a representation of the map; maps and territories are co-constructed . . . In other words . . . space is constituted through mapping practices, amongst many others, so that maps are not a reflection of the world, but a re-creation of it; mapping activates territory."[15]

The Streetmuseum app engages this tension between the representation of a space and the "territorialization" of that space by shifting between the top-down map of the London area and the image taken or created from a particular embodied perspective. For my inquiry into the uses of maps for locative media projects, I am curious about the relationship between these two levels of interface. Does the utilization of both the map and the highly particular embodied point of view (which can be experienced only if the user is standing in precisely the same spot that the photograph was originally taken) allow for a complementary relationship between the map and the image? Or does having users encounter a map first create a hierarchy in which the "objective" view allows for the "subjective"? One approach to answering these inquiries is to look at the ways that maps function as orienting objects.

OBJECTS OF ORIENTATION

A fruitful way of thinking about our relationship to maps is by looking at practices of orientation through a phenomenological lens. How do we get our bearings in a space? This is more than a matter of simply understanding your location; it is also a question about spatial relationships and how practices of being situated in—and moving through—a space are culturally inscribed. Scholars have investigated the ideas and practices of orientation in a wide range of ways, from Mark Hansen's work looking at the relationship between proprioception and affect,[16] Vivian Sobchack's investigation of the embodied experience of being lost,[17] and Sara Ahmed's theorization of orientation (including physical, experiential, and sexual orientations) via Husserl's "two-fold directedness."[18]

Our experience of implacement, as Edward Casey has termed it, is one of understanding our situational location.[19] This is done in a number of ways. We orient our bodies in a proprioceptive way (i.e., the sensation of understanding your body's positionality in relationship to the people and objects around you) through what I term a practice of "sensory-inscription."[20] To understand the power of maps in locative media projects, we have to understand that we orient ourselves in space (and thus understand our implacement) by simultaneously experiencing the space in a *sensory* way—as something that is learned as you move through it, engaging it with all of your senses—and understand that such implacement is culturally *inscribed* and contextually specific.

Ahmed's work on queer phenomenology demonstrates that objects (especially orienting objects like maps in locative media projects for my study) "arrive" to us to interact with them via networks of cultural and historical influences that determine their arrival (e.g., the shape such objects take and the anticipated ways in which we can interact with them). She writes, "What arrives not only depends on time, but is shaped by the conditions of arrival, by how it came to get here. Think of a sticky object; what it picks up on its surface 'shows' where it has traveled and what it has come into contact with. You bring your past encounters with you when you arrive."[21] The ways that we orient ourselves include not only interpersonal interactions with others nearby but also with objects in the space and objects that represent the space. The "arrival" of these objects as tools of orientation does not happen in a cultural vacuum; instead, our understanding of orientation via objects like maps comes packaged with a host of histories, both personal and historical, and with the forces of a wide range of cultural entities.

For example, when a person loads up a location-based service that uses a map as its primary interface—from Google Maps to Yelp, from Foursquare to Streetmuseum—the map centers on that person's location. He or she, in essence, becomes the center point around which the map pivots. The map orients to that person (in contrast to a shopping mall directory, as one example, which alerts people to their location through a symbol saying "You Are Here"). While this has been hailed as a dramatically different practice of orientation, it bears remarkable similarity to the ways that maps have been created throughout history. When a map like the sixteenth-century Mercator projection was created (primarily for nautical navigation across oceans), Europe was the center point for the map, and the northern hemisphere was given the top of the map. This was a practice that was carried out by each culture, situating itself as the center of the map. Thus, we have historically understood our position in space as central, with other objects, places, and people connecting to us from the center outward. Here, the practice of sensory inscription for orientation shows how important our maps have been for understanding our place in the world. Through this object, we have a sensory layering between physical space and the representation of that space. This helps us know our place in the world in a proprioceptive way that allows us to "see" our location in relationship to the local landscape that is out of sensory view. Simultaneously, we come to maps (and they arrive to us as tools of orientation) with a wide range of historical pressures and expectations. These cultural inscriptions on maps (which transform maps into inscriptions of culture) are read in a number of ways depending on the visibility of the inscriptions.

THE VISIBILITY OF MAPS

The various ways by which maps arrive to users in locative media projects serve either to obscure the sedimentation of historical and cultural forces that constrain how we interact with maps or, conversely, serve to expose those forces as a form of critique. In both cases, we must understand that—in line with Mark Monmonier's adage—all maps "lie." He argues:

> A good map tells a multitude of little white lies; it suppresses truth to help the user see what needs to be seen. Reality is three-dimensional, rich in detail, and far too factual to allow a complete yet uncluttered two-dimensional graphic scale model. Indeed, a map that did not generalize would be useless. But the value of a map depends on how well its generalized geometry and generalized content reflect a chosen aspect of reality.[22]

When utilizing a map in a project like Streetmuseum, this understanding of the maps as a host of white lies is likely *not* the way a user would describe the experience; instead, the map offers a limited set of data relevant to the purposes of the app. The map, that is, is "true" and "correct" based on the expectations a user brings to it (if the map within Streetmuseum works "correctly"). When engaging a locative media project that uses maps as the primary interface, we interact with the map in a way that reflects it as an "interfaceless interface," a term that Jay David Bolter and Richard Grusin use to describe the immediacy of a medium that delivers content without drawing attention to itself as a medium.[23] Indeed, maps tend to obscure their own authorship to deliver their content, thus seeming to erase the interface (and its politics) entirely. As a result, the networks of circulation that allow maps to arrive are often obscured.

Such obscuring of a map's networks of circulation results in spaces and media that address some bodies and not others. As Ahmed notes, the use of a map seems to be common sense: its use is inscribed in its very being. Objects like maps reveal their essence by what they allow people to do with them; when we look at a map in the Streetmuseum app, we know what it signifies and the various conventions of how to orient the map with our bodies and our bodies with the map. This commonsense approach to the perceived affordances of a map obscures the fact that, as Ahmed puts it, "an object tends toward some bodies more than others." She continues, "So it is not simply that some bodies and tools *happen* to generate specific actions. Objects, as well as spaces, are made for some kinds of bodies more than others."[24] Thus, when we ask the questions, "Whom does this map address, and whom does it exclude?" or "For whom is this locative media project intended?" we begin to uncover the fact that, as representations of our locations, maps present the space from particular perspectives and omit others. Maps thus tend to reiterate the practices and ideas about a space that reinforce the existing power dynamics of the space.

CRITIQUING THE MAP WITHIN THE MAP

The invisibility of these power dynamics in the maps that we use for locative media projects demonstrate that maps can be extraordinarily effectual tools for those in positions of power to maintain the status quo practices of the space by presenting them as common sense. As Raymond Craib has convincingly argued:

> [N]o other image has enjoyed such prestige of neutrality and objectivity [as the map has] . . . The most oppressive and dangerous of all cultural artifacts may be the ones so naturalized and presumably commonsensical as to avoid critique. Like any other production, a map is contingent on its sponsor and its producer and on their cultural, social, and political world desires.[25]

Many locative media artists and producers have utilized this relationship between objectivity and common sense as the launching pad for their own work. These artists have sought to critique the objectivity of maps by utilizing them in ways that question the very nature of mapmaking and cartography and instead insert the human body's subjective perspective as the primary way of seeing the world. These works range from Paula Levine's creation of maps that overlay distant places onto nearby ones (such as her layering of San Francisco and Baghdad during the first Gulf War) to Christian Nold's creation of "emotion maps" of various cities by having volunteers wander the space while he monitored their stress levels with a mobile galvanic skin response machine (i.e., a mobile lie detector that measured heart rate, perspiration, and breathing among other reactions). These projects seek to make the embodied perspective of individuals the primary way of encountering the map. Orienting yourself with these maps and navigating the space with them ultimately allows us as participants to question the very process of orientation and cartography.

Perhaps most notable is the work of Esther Polak, whose *Amsterdam REALtime* is widely regarded as one of the first forays into locative mapping. In this project, she equipped about 60 participants with GPS receivers and tracked their movements throughout the city. Instead of having an existing map as a base layer, she began with a black background that slowly transformed into the lived map of Amsterdam as people left behind traces of their pathways around the city over a period of six weeks. Her subsequent locative projects explored the relationship between visualization and embodied practices of space. As Lone Koefoed Hansen notes, Polak's recent projects, *Souvenir* and *NomadicMILK*, "address the difference between the person being tracked and the person trying to make sense of the tracks . . . [B]oth projects show how the tracked participants negotiate and navigate structures of politics, economy, and the concrete landscape through their movements."[26] Rather than mapping space (as a preexisting container), these projects map movement and subsequently create maps of the body's engagement with space.

The visualizations created in these projects often look very little like what we expect a map to look like. Creating maps of people's movements related to food and milk production in the Netherlands and Nigeria (the subjects and spaces of *Souvenir* and *NomadicMILK*), Polak's traces of tractor movements through a field or pathways of a nomadic dairy farmer in Nigeria produce images that do not serve as mimetic representations of the spaces. Instead, the traces created by the movements of these people create visualizations of the "trajectories [that] are experienced by the people making them. Using the GPS receiver as a storytelling tool, she accesses individual stories about everyday practices and how individuals experience their relationship to the larger structures that are part of their movements."[27] As Hansen argues, "I understand Polak's projects as site-specific narratives that both 'write' and 'read' a location with the heterogeneous 'voices' of those who embody the place."[28]

For emerging locative media projects, there is immense power in the ability to take something so familiar (and commonsense) as the map and get people to engage it in a new and unexpected way. Much of the power of locative media is found in the ways that these projects engage participants as embodied actors producing the space around them. When these participants are able to completely reimagine their relationship to how their spaces are visualized, the impact is profound: people are able to have critical distance on their own practices of space that have been taken for granted due to the invisibility of the map's networks of circulation. As such, locative designers, artists, and participants can take the map and creatively misuse it to design unexpected ways of visualizing (and thus practicing) the spaces they move through. This ultimately allows for a critique of the map within the map itself.

The firestorm that erupted immediately after the launch of Apple Maps, as discussed at the beginning of this chapter, brings to the fore these issues of a map's politics, power, and networks of circulation. The transformation of the default maps on the iPhone forced people out of an interfaceless-interface relationship to the maps that they used on a daily basis in order to orient themselves and navigate through the world. The maps they were presented with instead revealed the interconnection between maps and corporate interests. The maps also highlighted how indelibly linked our spatial practices are to the representations we use as ways of seeing the world. Engaging these maps critically allows for an analysis of the relationships among how we visualize the world, how we practice these spaces, and how these objects for orientation are also ways of orienting the self as a particular subject within the space. As locative media continues to draw on maps as a primary interface, while simultaneously situating the body's interactions in the world as the primary mode of producing space, these projects will engage us at the intersection of representation and practice, asking us to perform the tensions between our assumptions about our locations and our role in the production of space.

NOTES

1. Chris Taylor, "Apple Maps App: Just How Bad Is It?" *Mashable* (September 20, 2012), accessed August 15, 2013, http://mashable.com/2012/09/20/apple-maps-app/.
2. Anil Dash, "Who Benefits from iOS6's Crappy Maps?" *Anil Dash: A Blog About Making Culture* (September 19, 2012), accessed August 15, 2013, http://dashes.com/anil/2012/09/who-benefits-from-ios6s-crappy-maps.html.
3. Sarah Sharma, "It Changes Space and Time! Introducing Power-Chronography," in *Communication Matters: Materialist Approaches to Media, Mobility and Networks* (New York: Routledge Press, 2012), 66.
4. Sharma, "It Changes Space and Time!" 66–67.
5. Sharma, "It Changes Space and Time!" 67.
6. Henri Lefebvre, *The Production of Space*, trans. Donald Nicholson-Smith (Oxford: Blackwell Publishing, 1991), 83, 116.
7. Sharma, "It Changes Space and Time!" 67.
8. Lefebvre, *The Production of Space*, 89–90.
9. Jason Farman, *Mobile Interface Theory* (New York: Routledge, 2012), 35.
10. Charles Sanders Peirce, "New Elements," in *The Essential Peirce: Selected Philosophical Writings*, Vol. 2 (Bloomington: Indiana University Press, 1998), 322.
11. Mary Ann Doane, "Indexicality: Trace and Sign: Introduction," in *Differences: A Journal of Feminist Cultural Studies* 18, no. 1: 2.
12. Donna Haraway, *Simians, Cyborgs, and Women: The Reinvention of Nature* (New York: Routledge, 1991), 189.
13. See Jason Farman, "Mapping the Digital Empire: Google Earth and the Process of Postmodern Cartography," *New Media & Society* 12, no. 6 (2010): 869–888.
14. Jean Baurdrillard, *Simulacra and Simulation*, trans. S. F. Glaser (Ann Arbor: University of Michigan Press, 1994), 1.
15. Rob Kitchen, Martin Dodge, and Chris Perkins, "Introductory Essay: Conceptualising Mapping," in *The Map Reader*, eds. Rob Kitchen, Martin Dodge, and Chris Perkins (Oxford: Wiley-Blackwell Press, 2011), 6.
16. Mark B. N. Hansen, *New Philosophy for New Media* (Cambridge, MA: MIT Press, 2006).
17. Vivian Sobchack, "Breadcrumbs in the Forest: Three Meditations on Being Lost," in *Carnal Thoughts: Embodiment and Moving Image Culture* (Los Angeles: University of California Press, 2004).

18. Sara Ahmed, *Queer Phenomenology: Orientations, Objects, Others* (Durham, NC: Duke University Press, 2006).
19. Edward S. Casey, *Getting Back into Place: Toward a Renewed Understanding of the Place-World* (Bloomington: Indiana University Press, 2009), 36–37.
20. Farman, *Mobile Interface Theory.*
21. Ahmed, *Queer Phenomenology*, 40.
22. Mark Monmonier, *How to Lie with Maps* (Chicago: University of Chicago Press, 1996), 25.
23. Jay David Bolter and Richard Grusin, *Remediation* (Cambridge, MA: MIT Press, ?1999), 23.
24. Ahmed, *Queer Phenomenology*, 51.
25. Raymond B. Craib, "Cartography and Power in the Conquest and Creation of New Spain," *Latin American Research Review* 35, no. 1 (2000): 8.
26. Lone Koefoed Hansen, "Paths of Movement: Negotiating Spatial Narratives Through GPS Tracking," in *The Mobile Story: Narrative Practices with Locative Technologies*, ed. Jason Farman (New York: Routledge, 2013), 134.
27. Hansen, "Paths of Movement," 130.
28. Hansen, "Paths of Movement," 130.

7 Locating Media, Performing Spatiality

A Nonrepresentational Approach to Locative Media

Federica Timeto

INTRODUCTION

In the contemporary media environment, space and information do not stand still but are continuously enacted through the relations among subjects, objects, and places. For information and communication technologies (ICTs), particularly for mobile devices, the locational aggregation of data has become foundational,[1] increasing the availability, circulation, and acquisition of information. ICTs act as the interfaces that *mediate* the social and the spatial,[2] underlining their conjoined performativity.[3] Building on the mobile, networked approach that accounts for the so-called mobility turn of contemporary social practices,[4] challenging both the object of enquiry of the social sciences and existing research methodologies, I believe that we also need to question the representational paradigm itself—typically employed when relations between space and society are considered—in order to prioritize the *processes* and *relations* that characterize locative media today, without congealing them in a list of binaries suitable for representation.

Thus, in this chapter, I propose an approach that, dismantling some of the (often implicit) representational assumptions that have proven inadequate to account for the performative aspects of locative media, articulates representation differently.[5] For my considerations, I draw on the nonrepresentational turn that has emerged in recent decades in the field of human geography[6] and that transversally involves the social and hard sciences as well as the arts,[7] trying to reconnect bodies and geophysical worlds at the interface between matter and information.

I start by reconsidering the artistic genealogy of the definition of locative media so as to foreground the performativity of locative practices and the "creativity" of the material–informational continuum. Then, to explain how representational binaries work, I give some examples of the ways space and representation have been conflated while space has been put in opposition to a series of other concepts that are considered to be more dynamic, such as place and time, and devalued accordingly. I focus specifically on several examples of performative mapping not only because maps have been historically treated as privileged objects of representation in the visual arts[8] but also because, prior to the widespread accessibility of geographical information systems (GIS), mapping was primarily employed to support and "prove" representational thinking.

I continue by taking into account the fruitful consequences of adopting a relational notion of location[9] when approaching the "exceeding" performativity of locative media environments, which requires that we "address" rather than understand media and their networks of mediations[10] "not as if they were logical systems or structures but as if they were environments where images live, or personas and avatars that address us and can be addressed in turn."[11]

LOCATING MEDIA

All media can, in principle, be considered locative if we refer to the context in which they are produced and consumed. Locative media, however, make more processual ways of experiencing and performing space possible through their material (the possibility of transporting mobile devices inside and through physical places) as well as symbolic aspects (the possibility of moving through different forms of even virtual proximity). On the one hand, the mobility of locative media encompasses the possibility of producing and consuming information in movement. On the other hand, space, mobilized by information, can now be apprehended through its performativity.

Analogously, while the technical definition of locative media as media of communication that work through location-aware technologies, such as GPS, RFID, wireless networks, ubiquitous computers, smartphones, and wearables, clearly delineates a field of usage for locative devices, it does not necessarily imply a situated employment of such technologies.[12] A different, situated approach is required to account for the social practices of contextualization that take place in locative media environments, which presupposes a performative understanding of their mediations. This is even more the case because, in contemporary locative media, experiences appear to be not merely linked to physical location but to the hybrid "situation" of communication, a more complex but less binding form of positionality[13] that further entangles the already existing coimplications of reality and information.[14] Through locative media, the experience of being somewhere increasingly merges with media as the environment of experience, a phenomenon that the convergent outcomes of the spatial turn of media studies and the media turn of spatial disciplines also make visible.[15]

Although locative media are by now undoubtedly interdisciplinary,[16] keeping in mind the conjoined genealogy of locative media and locative arts can help us better approach the performative and relational qualities of the media environment without falling back on representational validation, along with the changes in the epistemological framework that this implies in both the social and the hard sciences.[17] One premise of such an approach, I think, is that we need to adopt a pragmatic and transversal notion of the esthetic, like the one proposed by Félix Guattari.[18] For Guattari, the esthetic is intended as a dimension of creation and virtuality that is not limited to the finite modes of artistic production but rather is concerned with the performative forces by which different space-times are accomplished and that "cannot be located in terms of extrinsic systems of reference" but only through "existential apprehension."[19] This idea points to a field of processes of differentiation that cannot be "circumscribed within the logic of discursive sets"[20] because their insides and outsides cannot be clearly delimited in advance, except through experiencing them in their becoming by means of an "ontological pragmatics."[21]

It is actually very difficult to talk about locative media without considering how they perform encounters, connections, and relations, which in turn recalls the impossibility of approaching locative media according to a representational framework that, on the contrary, works by identification and opposition, relying on a dimensional conception of space. Instead, as Richard Ek writes, "media and communication technologies are . . . primarily situated practices, verbs . . . manifesting the materialisation-and-symbolisation of human-and-material interaction through performances,"[22] a claim that is echoed by the artist Teri Rueb. Discussing her site-specific sound installation *Elsewhere/Anderswo* (2009), in which the participants activate "little elsewheres" in the landscape while walking with their positioning equipment, Rueb writes: "Place is a verb. Place making and the meaning of place, 'placings,' unfold as a continuous dialogue between the physical and built environment and its inhabitants. Landscape is a special kind of 'placing'."[23]

Let us also think about the locative artwork *Hertzian Rain*, by Mark Shepard,[24] in which audio streams are broadcast to a group of participants wearing wireless headphones and carrying umbrellas covered with electromagnetic field shielding fabric. Since the sound is transmitted wirelessly on the same local frequency, not only do the participants have to move to better receive the signals, but their personal spaces also create zones of interference with those of the others, who have to act accordingly. In turn, the resulting data is re-sent to the broadcasting station, which continuously modulates the sound emission. The space that the wireless networks create is thus highly unstable and circular, based on the contingency of situated connections, because after the very first sound transmission that activates the flow of communication, it then exclusively depends on the movements of the participants' bodies, the inclination of their umbrellas, and the unpredictable feedback loops these actions trigger in the sound environment.

An interesting parallel can be found in the way Adrian Mackenzie talks about wireless networks—the very infrastructure of networked media today—as the "prepositions ('at,' 'in,' 'with,' 'by,' 'between,' 'near,' etc.) in the grammar of contemporary media,"[25] and Karlis Kalnins's[26] idea of location as a "location of action or a time of the action," given that in many languages the locative case "corresponds vaguely to the preposition 'in,' 'at,' or 'by' of English."[27] As Mackenzie writes, the "conjunctive" experience of wireless networks "is diffuse, multiple and unstable in outline."[28] Each time wireless networks overflow, changing rate and direction, they create "equivocal proximities"[29] and forms of participation that do not depend on single subjects anymore but are instead disseminated along a network of shared and composite competencies.

A recent "dislocative" work by Julian Oliver, *Border Bumping* (2012),[30] is exemplary in this respect. A free downloadable app for smartphones, it allows the visualization of the signals that mobile devices send when they traverse territorial borders before the phones' owners do (Figure 7.1); these signals are transmitted to a server, which registers the trespassing "incidents" and maps an evolutionary telecartography of information whose unpredictability challenges the borders between states as well as those between subjects and objects (Figure 7.2). This points to a radically empirical notion of experience as "change taking place," at the same time *in situation* and *in transition*, that also contrasts with traditional notions of experience, assuming a "somewhat impersonal, pre-individual dimension"[31] in which relations have priority over actors.

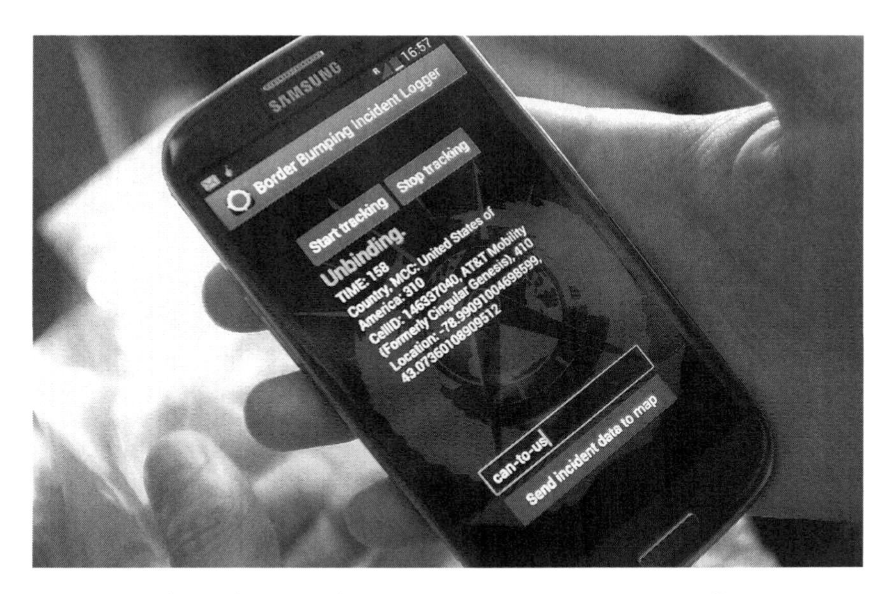

Figure 7.1 Julian Oliver, *Border Bumping* (2012–2013) (© Crystelle Vu, 2013).

Figure 7.2 Julian Oliver, *Border Bumping* (2012–2013) (© Julian Oliver, 2013).

ARTICULATING SPACE AND REPRESENTATION

When performing space,[32] locative media dismantle several related illusions: that places preexist their representations,[33] that an exact correspondence between a representation and the space represented exists, and that only one representation can be the objective one, acquiring an ideal status *per se*. They also combine reality and representation so that a whole series of mediations now involves the viewers, foregrounding their embodied—not merely visual—position.[34]

If space has traditionally been associated with representation, this is mostly because representation has commonly been thought of in terms of spatialization, as a way of fixing and stabilizing represented things in a given frame.[35] However, considering spatiality as a purified dimension, as a representable extension separated from the contingency of events, is no longer possible. The more the planes of our reality interface with one another through locative and ubiquitous networks, the more difficult it becomes to distinguish between places that are only physical and places that are only virtual, between materiality and information, between location and mobility. Being significantly "hybrid,"[36] locative media space cannot be, so to speak, purified. In fact, as Doreen Massey affirms, every representation is always a performing representation of a space-time,[37] enmeshed and embedded in its produced contingency.

This is not to say that space is impossible to represent; it is rather to say that it cannot be mimetically represented. In a traditional representational framework, the real world and the represented are considered two distinct realms; "reality," in consequence, turns out to be what could actually be properly represented as an object for the viewing subject. The gap postulated between the act of representing and the objects to be represented resolves the issue of the accuracy of representations by means of correspondence between the copy and the model. This move, however, also congeals both representation and space, disguising both their dynamism and their conjoined production. On the contrary, to mobilize space together with its representations, representation needs reworking as "an element in a continuous production" rather than "a process of fixing,"[38] so that the complication between the subject and the object of representation eventually emerges.

Space and representation have been traditionally conflated in many ways along a series of conceptual oppositions that are devoid of both mobility and ultimately reciprocity. For instance, there are the well-known oppositions between space and place and between place and time, which, equating space with stasis and closure, have reinforced the association between spatiality and representation, given that space has been accused of being anti–temporal because of its representational character.[39] The distinction between real and digital spaces also relies on a traditional representational framework[40] that, opposing the "here" of space and the "there" of media and information technologies, ultimately impedes a relational and nonrepresentational approach to both space and media.[41]

The idea that we enter a whole new territory because of digital flows, as well as the attempt to delineate its new spatial features, is part of an enduring belief in the autonomous existence of the *res extensor*: a construction that often relies on a mimetic interpretation of mapping techniques resulting in the complete erasure of material networks and the technologies through which representations are produced and producing.[42] The underlying assumption of a representational employment of maps is that they represent the real, even when accompanied by the paradoxical understanding that reality can ultimately be accessed only through representations.[43] Representational cartography has used the map in the same way that the Shannon–Weaver theory intended communication, that is, according to an idea of the reduction of the error and distortion necessary for the optimization of data transmission.[44] Space has accordingly been treated as a passive surface, a vehicle or receptacle for information.[45]

However, since the mid-1980s, thanks to changes introduced in computer graphics, the emergence of digital revisualization, and critical cartography,[46] the concept and employment of maps have changed, moving away from their treatment as simple products and toward their progressive consideration in terms of practices, requiring an always more interactive participation in and shared construction of space. This is particularly evident in the various forms of geoannotation and geotagging employed with both artistic and commercial aims in locative media,[47] from the pioneering experiments of the Proboscis collective, such as the renowned *Urban Tapestries* (2002–2004), to applications such as the user-generated architecture guide *Mimoa* (2007)[48] and the historical archive of personal and collective memories, *Historypin* (2010),[49] to name only two sets of examples, in which the mobile experience of space becomes the hybrid and always reversible result of both material and informational accessibility.

Postrepresentational cartography[50] considers the map as an interface not only of visualizing but of actually entering a network that does not stand in front of us but that is the location of everyday life.[51] In performative mapping practices, space is figured and practiced at the same time, revealing the coconstitution of space and maps and an implosion of the barrier between the cognitive apprehension and the embodied practice of space. Many recent "mapping performances," such as community-oriented actions like *Street with a View*, organized by Robin Hewlett and Ben Kinsley (2008) in Pittsburgh, or *What If Google Maps Went Live* (2011), by Studio Moniker in Eindhoven, as well as Aram Bartholl's urban interventions, including his Google Street View self-portrait series in Berlin, *15 Seconds of Fame* (2010), are cartographic processes that literally embody mobile annotation, making the data and the practices of space converge.

Performing place, here, becomes an "exceeding" relational activity that, taking the subjects and the objects of representation *elsewhere*, discloses the creativity of the locative media environment that pertains to both matter and information. This is the generative force that Mackenzie, drawing on Gilbert Simondon, has defined as "transduction" in his nonrepresentational account of technologies.[52] The properties of technologies as tools, for Mackenzie, are less important than their performing "capacity" as *events* to connect and produce new relations.[53] "Transduction" is precisely this modulation of forces that inform the contingent operations of a network, alternatively transforming and stabilizing it.[54] Accordingly, space is transducted, occurring as code/space, as Kitchin and Dodge (drawing on Mackenzie) write, every time it is modulated by information, as a "set of unfolding practices that lack a secure ontology—rather than a container or a plane of predetermined social production that is ontologically fixed."[55] This, in turn, discloses the relationality of the social in its mutual constitution with code/space, as the space where software and spatiality are "produced through one another."[56]

EXCEEDING LOCATIONS

The space of mobile and locative media is experienced not as an existing extension delimited by fixed entrances and exits but as an unfolding environment traversed by mobile interfaces that actively mediate the networks of sociospatial relations, highlighting their local–global continuum. As Ek argues,[57] relational thinking about space and the replacement of a dualistic framework with a dialectical one is the most suitable way to discuss mobile media: in fact, it exposes the inappropriateness of a dimensional and scalar logic of space and instead privileges connectivity and virtual articulation as the most relevant affordances of contemporary media-spaces.

As actor–network theory in particular shows,[58] the local is networked as well as mediated, generated, and distributed through a chain of associations that the ubiquity of the material infrastructure makes more traceable. Location continuously exceeds its *whereness* without, however, escaping it. Outside a local, there happens to be another local,[59] so that mobility articulates the local rather than being an opposite condition. When space can be elsewhere, exceeding its givenness, it can also be otherwise. In this tension, the actuality of location, being processual and not static, also reveals its virtuality—foregrounding the creative quality of space in the aforementioned sense: for each practiced space-time, a series of other possible space-times exist.[60]

This exceeding of location has been specifically theorized in the scholarship of locative media as "network locality"[61] or "net locality."[62] Mizuko Ito defines it as an "unbounded and dynamic . . . ongoing partial achievement,"[63] whereas Gordon and de Souza e Silva use the term to imply a more general shift in the way society and technology reciprocally respond to a growing need for mobile contexts, in which "the radical visibility of located data creates the *potentiality* for users to experience meaningful nearness to things and people."[64] For these reasons, approaching the media environment in terms of a sociotechnical framework, J. W. T. Mitchell affirms that media should be "addressed" rather than understood.[65] In fact, whereas understanding presupposes a subject that relates to an external object through cognitive apprehension, as in the representational framework, the idea of addressing situates the subject next to the object—an affirmation that appears particularly relevant when we talk about locative media that specifically revolve around the "sendings and receivings of socio-technical life," to use Nigel Thrift's expression.[66]

According to Thrift, the vast expansion of new modes of "readdressing the world"[67] in recent decades, from the track-and-trace model to the diffusion of spatiotemporal standardization, archiving and classification systems, logistics and demography, has contributed to the creation of a "continuous informational ethology"[68] in which the devices themselves have become "able to interact, dialogue, and adapt to users and other devices,"[69] acting inside a medium that "is constituted by a history of practices, rituals and habits, skills and techniques, as well as by a set of material objects and spaces."[70] Somehow paradoxically, says Mitchell, even though media do not have a specific address, they do contain all the possibilities of addressing, being the context in which "we live and move and have our being."[71] Thus, addressing does not belong to someone or something in particular; it is a capacity of the environment rather than a property of its actors.

Expanding the environmental definition of media,[72] Mitchell overtly blurs any binary used to define media, drawing attention to the *mediations* of the environment. Significantly, he further defines media as "ever-elastic middles that expand to include what look at first like their outer boundaries. The medium does not lie between sender and receiver; it includes and constitutes them."[73] Considering media as "middles," that is, as mediating environments that actively perform relations that in turn create complex social connections rather than solely as vehicles of information, means conceiving mediation as the complex of activities and forces through which the elements composing the media environment, be they social, technical, or spatial, find common actualization.

There is a big difference when the media through which the social is produced, including the media of communication, are treated as intermediaries or mediators, explains Latour.[74] As in the transmissive model of information, intermediaries transport something from one point to another without altering what they transport, so that the inputs and outputs eventually coincide. This kind of relation foresees no actual mediation, in Latour's sense, because the related elements remain extrinsic to one another. On the contrary, mediators, according to Latour, "transform, translate, distort and modify the meaning or the elements that they are supposed to carry,"[75] which implies a kind of relation that "does not transport causality but induces two mediators into coexisting."[76] As Mackenzie also observes about wireless networks:

> many of these changes seek to connect or align what was previously separated or misaligned. But in almost every attempt to converge, they disturb the rankings of conjunctive relations between impersonal and personal, between remote and intimate. The flow of experience has to be re-configured.[77]

In the field of locative media, objects become simultaneously at-hand and expanding environments that "make relational demands associated with their expandability."[78] Mediation, then, always encompasses an excess, a yet-to-come that cannot be deduced in advance. Analogously, although repetition and automatism make locative media reliable, nonetheless they are not *expressions* but *performances* of the material infrastructure, which also always comprise instability. In fact, if talking about expression presupposes a subject of representation, as well as a priority of representation, the notion of performance does not rely on any foundational premise of this sort, dislocating subjects as well as representations toward a "coalitional assemblage" that cannot "be known prior to its achievement."[79]

CONCLUSIONS

Locative media, specifically thematizing spatiality as communication, are *addressing media* in a space that does not preexist its relations but that is continuously performed as well as mediated. The diffuse addressability of locative media implies a performative redefinition of society, technology, and information, and it also requires a different approach that does not rely on external systems of validation but that is itself based on our "mutual and reciprocal constitution" with media.[80]

As I have argued in this chapter, the most important consequence of adopting a nonrepresentational approach to locative media—one that draws on the artistic origin of locative practices and focuses on networks, connections, and interactions rather than on boundaries, dimensions, and distinctions—is the rearticulation of space and representation: the former comes to be seen as a heterogeneous domain of relations that require continuous engagement, and the latter becomes a situated practice that does not depict reality from afar but that contributes to its construction and transduction from within. Accordingly, the social becomes reconceived as a distributed field of agencies whose beings are locally and contingently performed through the informational ethology of the media environment, whose addressability and "creativity" allow for situated and networked experiences to *take place* somewhere, while also taking *place* elsewhere.

NOTES

1. Eric Gordon and Adriana de Souza e Silva, *Net Locality. Why Location Matters in a Networked World* (Malden, MA: Wiley Blackwell, 2011), 11.
2. Bruno Latour, *Reassembling the Social: An Introduction to Actor-Network-Theory* (Oxford: Oxford University Press, 2005); Adriana de Souza e Silva, "From Cyber to Hybrid: Mobile Technologies as Interfaces of Hybrid Spaces," *Space & Culture* 9, no. 3 (2006): 261–278; Jason Farman, *Mobile Interface Theory: Embodied Space and Locative Media* (London: Routledge, 2012).
3. Adrian Mackenzie, "Transduction: Invention, Innovation and Collective Life" (2003), accessed May 28, 2013, www.lancs.ac.uk/staff/mackenza/papers/transduction.pdf; Nigel Thrift, *Non-Representational Theory: Space/Politics/Affect* (London: Routledge, 2008).
4. Mimi Sheller and John Urry, "The New Mobilities Paradigm," *Environment and Planning A* 38, no. 2 (2006): 207–226.
5. Thrift, *Non-Representational Theory*; Andrew Pickering, "After Representation: Science Studies in the Performative Idiom," Proceedings of the Biennial Meeting of the Philosophy of Science Association 2 (1994): 413–419; Federica Timeto, "Diffracting the Rays of Technoscience: A Situated Critique of Representation," *Poiesis & Praxis* 8 (2012): 151–167.
6. Nicky Gregson and Gillian Rose, "Taking Butler Elsewhere: Performativities, Spatialities and Subjectivities," *Environment and Planning D* 18, no. 4 (2000): 433–452; Hayden Lorimer, "Cultural Geography: The Business of Being 'More-Than-Representational,'" *Progress in Human Geography* 29, no. 1 (2005): 83–94; Sarah Whatmore, "Materialist Returns: Practising Cultural Geography in and for a More-Than-Human World," *Cultural Geography* 13, no. 4 (2006): 600–609.
7. See Thrift, *Non-Representational Theory*, and Timeto, "Diffracting the Rays," for further references and a literature review.

8. Ronald Rees, "Historical Links Between Cartography and Art," *Geographical Review* 70, no. 1 (1980): 60–78.
9. Richard Ek, "Media Studies, Geographical Imaginations and Relational Space," in *Geographies of Communication*, eds. Jesper Falkheimer and André Jansson (Göteborg: Nordicom, 2006), 43–64; Mizuko Ito, "Network Localities: Identity, Place and Digital Media" (paper presented at Meetings of the Society for the Social Studies of Science, San Diego, CA, 1999), accessed June 1, 2013, www.itofisher.com/PEOPLE/mito/papers/locality.pdf); Gordon and de Souza e Silva, *Net Locality.*
10. Latour, *Reassembling the Social.*
11. J. W. T. Mitchell, "Addressing Media," *MediaTropes* 1 (2008), 3, accessed June 1, 2013, www.mediatropes.com/index.php/Mediatropes/article/view Article/177.
12. See, for example, Jordan Crandall, "Precision + Guided + Seeing," *CTheory 1000 Days of Theory* (2006), accessed June 1, 2013, www.ctheory.net/articles.aspx?id=502; Francesco Lapenta, "Geomedia: On Location-Based Media, the Changing Status of Collective Image Production and the Emergence of Social Navigation Systems," *Visual Studies* 26, no. 1 (2011): 14–24; Andrea Zeffiro, "The Persistence of Surveillance: The Panoptic Potential of Locative Media," *Wi: Journal of the Mobile Digital Commons Network* 1 (2006), accessed May 28, 2013, http://wi.hexagram.ca/1_1_html/1_1_zeffiro.html.
13. de Souza e Silva, "From Cyber to Hybrid"; Joshua Meyrowitz, "The Rise of Glocality: New Senses of Place and Identity in the Global Village," in *A Sense of Place: The Global and the Local in Mobile Communication*, ed. Kristóf Nyíri (Vienna: Passagen Verlag, 2005); Rowan Wilken, "From *Stabilitas Loci* to *Mobilitas Loci*: Networked Mobility and the Transformation of Place," *The Fibreculture Journal* 6 (2005), accessed May 28, 2013, http://six.fibreculturejournal.org/fcj-036-from-stabilitas-loci-to-mobilitas-loci-networked-mobility-and-the-transformation-of-place/.
14. Keller Easterling, "An Internet of Things," *e-flux* 31, no. 1 (2012), accessed May 28, 2013, www.e-flux.com/journal/an-internet-of-things/; André Lemos, "Mobile Communication and New Sense of Places: A Critique of Spatialization in Cyberculture," *Revista Galáxia* 16 (2008): 91–108.
15. Ek, *Media Studies*, 54; Tristan Thielmann, "Locative Media and Mediated Localities: An Introduction to Media Geographies," *Aether. The Journal of Media Geography* 5A (2010): 1–17.
16. Rowan Wilken, "Locative Media: From Specialized Preoccupation to Mainstream Fascination," *Convergence* 18, no. 3 (2012): 243–248.
17. Marc Tuters, "Locating Locative: The Genealogy of a Keyword," *Acoustic Space* 10 (2011).
18. Félix Guattari, *Chaosmosis*: *An Ethico-Aesthetic Paradigm*, trans. Paul Bains and Julian Pefanis (Sydney: Power Publications: 1995); Adriana de Souza e Silva and Daniel M. Sutko, "Theorizing Locative Technologies Through Philosophies of the Virtual," *Communication Theory* 21, no.1 (2011): 23–42.
19. Guattari, *Chaosmosis*, 92–93.
20. Guattari, *Chaosmosis*, 92.
21. Guattari, *Chaosmosis*, 94. Interestingly, Anna Munster, who proposes a baroque interpretation of digital machines, reads new media and information aesthetics within a field of differential forces in which all the binaries that have characterized the popular narratives of digital media impinge on one another rather than excluding one other. In this context, she defines the virtual as "a set of potential movements produced by forces that differentially work though matter, resulting in the actualizations of that matter under local conditions"; see Anna Munster, *Materializing New Media: Embodiment in Information Aesthetics* (Lebanon, NH: Dartmouth College Press, 2006): 90.

22. Ek, *Media Studies*, 54.
23. Teri Rueb, *Elsewhere/Anderswo* (2009), accessed September 17, 2013, www.terirueb.net/elsewhere/.
24. Mark Shepard, *Hertzian Rain* (2009), accessed September 17, 2013, www.andinc.org/v3/hertzianrain/.
25. Adrian Mackenzie, "Wirelessness as Experience of Transition," *Fibreculture* 13 (2008), accessed May 28, 2013, http://journal.fibreculture.org/issue13/issue13_mackenzie_print.html#top.
26. Kalnins is believed to be the first to connect the notion of locative media to artistic research in the field of locative technologies during the Art+Communication Festival in Riga, Latvia, in 2003. For a different view, however, see Andrea Zeffiro, "A Location of One's Own: A Genealogy of Locative Media," *Convergence* 18, no. 3 (2012): 249–266.
27. Karlis Kalnins, communication to *Locative Mailing List*, "Locative Is a Case Not a Place" (May 10, 2004), accessed May 30, 2013, http://web.archive.org/web/20050907080544///db.x-i.net/locative/2004/000385.html.
28. Mackenzie, "Wirelessness as Experience."
29. Mackenzie, "Wirelessness as Experience."
30. Julian Oliver, *Border Bumping* (2012), accessed May 30, 2013, http://borderbumping.net/.
31. Oliver, *Border Bumping.*
32. Which, of course, is not always the case inasmuch as performativity is an affordance rather than a property of locative media. See Lapenta, "Geomedia: On Location-Based Media."
33. See Edward S. Casey, *The Fate of Place* (Berkeley: University of California Press, 1998); Jeff Malpas, *Place and Experience: A Philosophical Topography* (Cambridge: Cambridge University Press, 1999).
34. See Martin Rieser, "Beyond Mapping: New Strategies for Meaning in Locative Artworks," in *The Mobile Audience: Media Art and Mobile Technologies*, ed. Martin Rieser (Amsterdam: Rodopi, 2011).
35. Doreen Massey, *For Space* (London: Sage, 2005).
36. de Souza e Silva, "From Cyber to Hybrid."
37. Massey, *For Space*, 27.
38. Massey, *For Space*, 28.
39. See Massey, *For Space*, 45 *ff*., for a more detailed critique of theorists with nondynamic concepts of space.
40. Richard Coyne, *Technoromanticism: Digital Narrative, Holism, and the Romance of the Real* (Cambridge, MA: MIT Press, 1999).
41. Ek, *Media Studies, 51 ff.*
42. Valérie November, Eduardo Camacho-Hübner, and Bruno Latour, "Entering a Risky Territory: Space in the Age of Digital Navigation," *Environment and Planning D* 28, no. 4 (2010): 581–599.
43. The genealogy of such a supposition can be traced back to the role of linear perspective and down to the invention of the *camera obscura*, reaching its apotheosis around the middle of the twentieth century when geography was finally constituted as a formal spatial science. See Derek Gregory, *Geographical Imaginations* (Cambridge, MA: Blackwell, 1994); Rees, "Historical Links."
44. Rob Kitchin, Chris Perkins, and Martin Dodge, "Rethinking About Maps," in *Rethinking Maps*, eds. Rob Kitchin, Chris Perkins, and Martin Dodge (London: Routledge, 2009), 1–25.
45. Mike Crang and Stephen Graham, "Sentient Cities: Ambient Intelligence and the Politics of Urban Space," *Information, Communication & Society* 10, no. 6 (2007): 789–817; Federica Timeto, "Redefining the City through Social Software," *First Monday* 18, no. 11 (2013), accessed May 28, 2013, http://journals.uic.edu/ojs/index.php/fm/article/view/4952.

46. Jeremy W. Crampton and John Krygier, "An Introduction to Critical Cartography," *ACME Journal* 4, no. 1 (2006): 11–33.
47. Gordon and de Souza e Silva, *Net Locality*, 40 *ff.*
48. MIMOA, *mi modern architecture*, accessed May 30, 2013, http://www.mimoa.eu.
49. Historypin, accessed May 30, 2013, http://www.historypin.com.
50. Kitchin, Perkins, and Dodge, "Rethinking About Maps."
51. Eric Gordon, "Mapping Digital Networks from Cyberspace to Google," *Information, Communication & Society* 10, no. 6 (2007): 885–901.
52. Mackenzie, "Transduction."
53. Mackenzie, "Transduction," 19.
54. Mackenzie, "Transduction," 10.
55. Rob Kitchin and Martin Dodge, *Code/Space: Software and Everyday Life* (Cambridge, MA: The MIT Press, 2011), 16.
56. Kitchin and Dodge, *Code/Space.*
57. Ek, *Media Studies.*
58. Latour, *Reassembling the Social.*
59. Latour, *Reassembling the Social*, 191 *ff.*
60. Thrift, *Non-Representational Theory*, 120–121.
61. Ito, "Network Localities."
62. Gordon and de Souza e Silva, *Net Locality.*
63. Ito, "Network Localities," 21.
64. Gordon and de Souza e Silva, *Net Locality*, 12–13.
65. Mitchell, "Addressing Media."
66. Nigel Thrift, "Remembering the Technological Unconscious by Foregrounding Knowledges of Position," *Environment and Planning D* 22, no. 1 (2004), 175.
67. Thrift, "Remembering," 182.
68. Thrift, "Remembering," 185.
69. Thrift, "Remembering," 183.
70. Mitchell, "Addressing Media," 12.
71. Mitchell, "Addressing Media."
72. See Marshall McLuhan, *The Gutenberg Galaxy: The Making of Typographic Man* (Toronto: University of Toronto Press, 1965); Joshua Meyrowitz, *No Sense of Place: The Impact of Electronic Media on Social Behavior* (New York: Oxford University Press, 1985).
73. Mitchell, "Addressing Media," 4.
74. Latour, *Reassembling the Social*, 37 ff.
75. Latour, *Reassembling the Social*, 39.
76. Latour, *Reassembling the Social*, 108.
77. Mackenzie, "Wirelessness as Experience."
78. Karin Knorr-Cetina, "Postsocial Relations: Theorizing Sociality in a Postsocial Environment," in *Handbook of Social Theory*, ed. George Ritzer and Barry Smart (London: Sage, 2001), 530. To highlight this relational capacity of the contemporary media environment, Latour defines "plug-ins" as the "subjectifying" components that reside neither in subjects nor in objects but that allow social actors to "*become* locally and provisionally competent." See Latour, *Reassembling the Social*, 207.
79. These observations by Judith Butler regard the performance of gender. See Butler, *Gender Trouble: Feminism and the Subversion of Identity* (1999, London: Routledge), 21. Butler's notion of performativity has been subsequently employed in nonrepresentational geography, as in Gregson and Rose, *Taking Butler Elsewhere.*
80. Mitchell, "Addressing Media," 14.

8 The Cluster Diagram
A Topological Analysis of Locative Networking

Carlos Barreneche

"Coded spaces"[1] are transforming the nature of social space[2] and, as a result, the space within which "class, culture and identities play out."[3] I have argued elsewhere that locative media represent not a primarily geographical mediation of the city but rather a geodemographical mediation, for at play is the sorting of places according to social relations.[4] In this chapter, I want to further extend that thesis by arguing that an elementary geodemographic logic is in fact inscribed at the very core of the computational processes of ordering subjacent to locative networking.

Geodemographic information systems (GDIS) are built on the sociological assumption that location, particularly where we live, signals the social and cultural characteristics of a given population. Contemporary commercial locative media platforms introduce a different geodemographic framework. Whereas traditional geodemographics are built on the category of residency (postal codes), in these platforms households are not the basic unit of calculation but "places" or the so-called "venues"—categorized at the ontological level of the database as points of interest (POI).

In order to analyze the geodemographic spatial rationality underpinning locative media systems, I will use a model proposed by Noulas, Scellato, Mascolo, and Pontil.[5] This model is not a description of any of the current location-based services; nevertheless, it could be considered paradigmatic inasmuch as it concretizes what I want to argue are their core geodemographic logics. In a paper entitled "Exploiting Semantic Annotations for Clustering Geographic Areas and Users," the authors present a model that uses places metadata extracted from a social location platform (Foursquare) in order to produce profiles of geographical areas and human activity. The methodology adopts the same platform's place categorizations (e.g., food, nightlife, shops, home, etc.) to create a profile of a given area based on the aggregate of places (POI) that such an area contains and their respective annotations (place metadata). The computer scientists explain the procedure in more detail:

> We consider a centre point g within a city and a large square area A. We split A into a number of equally sized squares, each one representing

> a smaller local area *a*. Each area *a* will be a data point input for the clustering algorithm. The representation of *a* is defined according to the categories of *nearby places* and the attached *social* activity modelled through the number of checkins that took place at those. In this way not only do we know what types of places are in an area, but we also have a measure of their importance from a social point of view.[6]

Similarly, user profiling is achieved by way of linking people with places through check-ins and correlating them against the places' respective categories, thus allowing for the identification of patterns of co-visitation, that is, clusters of users that visit similar categories of places (e.g. people who go to X tend to go to Y).[7] We can extrapolate a generalizable principle from this model for an understanding of the geodemographic ontology of locative media: geographical areas may be profiled according to the aggregate of nearby places (POI) and their respective annotations (place metadata). Equally, people may be profiled in terms of their aggregate relation to places (POI). This principle contrasts with GDIS in which neighborhood lifestyle segmentations are formed based on the demographics of the area's inhabitants. The very same geodemographic rationality—"identify and describe people by place, and vice versa"[8]—is actualized through the technological affordances of our current media-technological system: mobility, participation, and real time.

Mobile location-enabled devices and place databases have permitted the recording and sharing of location information and its subsequent data mining for mobility patterns, thus replacing the household as the unit of measurement for the POI. In consequence, if traditional GDIS geodemographic bias assumed that social identity corresponds to residency (i.e., similar people live in similar places), under the locative media guise, the geodemographic bias assumes that social identity corresponds to spatial mobility (i.e., similar people visit similar places). First GDIS had to assemble data collected from credit bureaus and commercial mailing lists in order to track people's mobility.[9] Current location platforms can register mobility in real time, so the system is fed back continuously with every new user-contributed annotation.

THE TOPOLOGICAL COMPUTATION OF SOCIAL SPACE

To understand further the geodemographic logic subjacent to present locative media, I will borrow the concept of the "diagram" from Deleuze. In Deleuze's words, the diagram is an "abstract machine," "a functioning," which "must be detached from any specific use."[10] The diagram, Deleuze explains, "has nothing to do either with a transcendent idea or with an ideological superstructure, or even with an economic infrastructure"; nonetheless it "acts as a non-unifying immanent cause."[11] The working hypothesis is that the form that sociospatial connections assume in locative media is characterized by proximity, and its corresponding topological pattern is the cluster (or clustered network). For the purpose of the analysis, I will refer to this hypothesis as the "cluster diagram."

Clustering classification is by no means exclusive of GDIS. As a matter of fact, it is pervasive in our present digital networked media. This mode of classification is constituent of recommender systems (or social-filtering systems) that today shape our cultural consumption (e.g., Amazon for books, Netflix for video, and Pandora for music). Likewise, clustering methods are central in social media to map and suggest networks of contacts/friends (e.g., Facebook or LinkedIn). Online advertising, particularly contextual advertising and behavioral targeting, is also based on clustering classification.

The majority of traditional GDIS use *k*-means clustering as a means to construct its classifications.[12] Even though GDIS providers keep these methodologies obscured for commercial reasons, Harris, Sleight, and Webber provide a description of the standard procedure.[13] The basic task is to aggregate large data sets into small units or clusters. *K*-means algorithms use a deductive model approach to achieve this goal. Firstly, the classifiers choose an initial number of clusters (k), often based on variables that signal types of consumers, although they can also be randomly selected by the algorithm. Then each neighborhood is allocated into a single cluster based on its proximity to each cluster means. Because this is an iterative method, the clustering process is repeated and zip codes are reallocated when necessary until the clusters become stable.[14] "In computational terms, the classification process is usually devised to maximize within cluster homogeneity while maintaining heterogeneity between clusters."[15]

The clustering classification technique predominant in locative media systems is the *k*–nearest neighbor method. In contrast to the *k*-means method, this is an inductive method whereby an unclassified object is not sorted according to a set of predefined categories but is classified based on the dominant characteristics of its neighbors. "It assigns to an unclassified point the class most heavily represented among its k, nearest neighbors,"[16] where *k* corresponds to the number of neighbors considered in the calculation. The *k*–nearest neighbor classification might be implemented in different forms and hybrid approaches to classification in location-based services. Nonetheless, what I want to draw out here for the purpose of the analysis is the common core computational primitive, the so-called nearest neighbor rule (NN rule). The NN rule was first proposed by Fix and Hodges in 1951,[17] and it can be bluntly formulated as the classification of an object (point) according to the majority vote of its nearest objects (neighbors) (Figure 8.1). The NN rule is foremost a measure of similarity by adjacency relations. Extrapolating from the technical definition, it can be argued that nearest-neighbor classification embodies the fundamental geodemographic principle that equates vicinity (similar neighborhood) with identity (similar demographics) because it assumes that closer objects in space are more related to

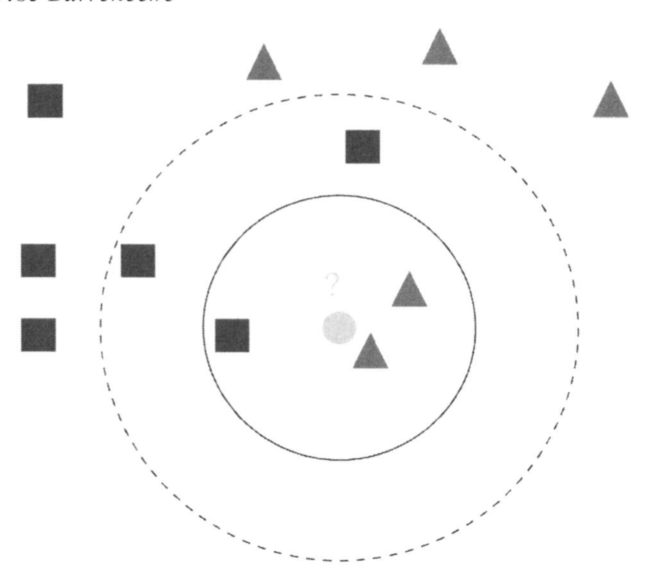

Figure 8.1 Nearest-neighbor classification diagram (source: "*k*-nearest neighbor algorithm," Wikipedia, May 28, 2007; upload by Antti Ajanki).

each other than distant objects are, so they are more likely to belong to the same class.

Interestingly, the NN rule also mirrors Tobler's first law of geography, which states that "everything is related to everything else, but near things are more related than distant things."[18] In fact, the NN rule handles data spatially. It assumes that data consists of points in space (feature space). This way, the NN rule maps spatial relations among points, distance, and neighborhood. However, despite this spatial bias, nearest-neighbor computation may also include the calculation of non-Euclidean distance metrics, like similarity measures.

Unlike GDIS fixed segmentation models, in locative media platforms, geodemographic sorting works on metastable orderings (clusters of people and places) calculated by machine learning algorithms. The abstract space at work is fundamentally different. The spaces of GDIS are fixed topographic grids. The abstract space corresponding to the network of relations between places and users abstracted in locative media's underlying databases, the so-called place graph,[19] on the other hand, is basically topological, that is, a fluctuating field of relations. Unlike topography, Mol and Law argue:

> Topology doesn't localize objects in terms of a given set of coordinates. Instead, it articulates different rules for localizing in a variety of coordinate systems. Thus it doesn't limit itself to the three standard axes, X, Y and Z, but invents alternative systems of axes.[20]

Accordingly, in locative media platforms, space is not only measured in terms of metric distance but also articulates alternative metrics. To illustrate, from a local search (e.g., Google Maps) to social location services (e.g., Foursquare), spatial entities are not sorted based merely on geographical distance. The nearest point is not necessarily the closest in Euclidean space (distance relations). In the topological space of the place graph, two physically distant points can be closely connected through social affinity (topological relations). Distant areas may share a similar geodemographic profile. Even places could be matched on a global scale. For example, an application may find the equivalent of London's Soho in Rome or may recommend to tourists what places to visit in a city based on the preferences from other cities they might have already visited. The place graph, therefore, positions us in geographical and social space at the same time.

The NN rule is especially central in the ubiquitous information milieu of locative media. For instance, it is at work in mobile local search whenever there is a query for finding the objects closest to a specified location (nearest neighbor search) (e.g., find the nearest café). The NN majority voting logic serves as a measurement of value for places in local search. Google's local algorithm (PlaceRank), for instance, measures the density of nearest neighbors in order to determine relevance.[21] Moreover, location-aware recommender systems—integral to social location platforms—may use the nearest-neighbor model to determine similarities between places and between places and users to suggest destinations. This is the specific case of Foursquare, for instance.[22]

As a matter of fact, the nearest-neighbor problem is the central operative logic in classification and search in various other areas. Examples include pattern recognition, genetic algorithms (gene expression analysis), and recommender systems. The NN rule appears central to our current informational condition characterized by the overabundance of data. Granted that, I want to put forward the hypothesis that the NN rule is nonetheless a primordial principle of organization in locative media. In other words, the NN rule would embody a topological invariant of ordering, the cluster form, according to which social space is modeled in these systems.

In a generic description, the NN rule would organize the place graph by way of tracing out patterns or relations of proximity among nodes. The place graph is partitioned in the process into clusters of proximity/similarity nodes. A clustered network form is configured so that all nodes (places and users) are assigned to a connected cluster or "neighborhood" (Figure 8.2). The NN rule thus engineers connections between places and users, introducing stratification in an initially distributed network. In this sense, it works as a "topological machine."[23]

Proximity—the nearest-neighbor relation, as it were—is the ordering logic of this diagram. However, this concept goes beyond geographic proximity to include also social proximity, that is to say, similarity (e.g., affiliation, social relevance, shared consumption patterns, etc.). A case in point

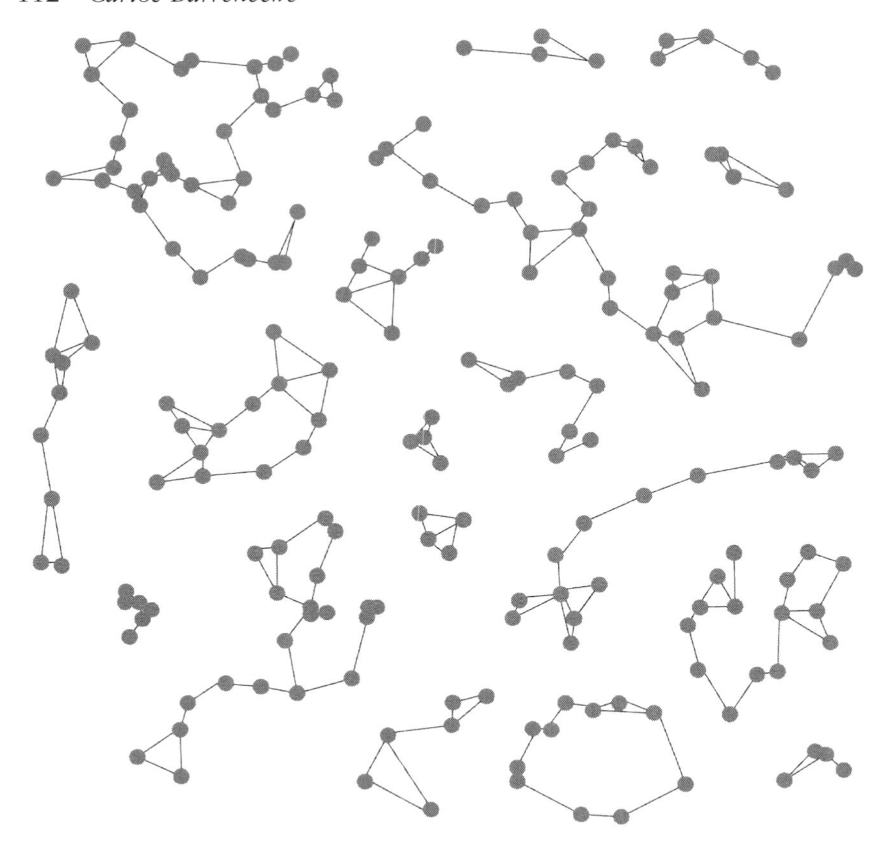

Figure 8.2 Nearest-neighbor network topology (source: "Nearest Neighbor Networks," Wolfram Demonstrations Project, accessed December 10, 2013. www.demonstrations.wolfram.com/NearestNeighborNetworks/).

is that social location platforms monitor users' spatial behaviors and social interactions to find regularities and matches of the kind "people who go to X tend to go Y" (e.g., Foursquare and Facebook Places). Because proximity is both geographical and social, locability and identifiability are here equated. In other words, in the cluster diagram, the user is located both in space and in social space at the same time, enacting thus a geodemographic ontology.

THE AUTOMATIC PRODUCTION OF THE CITY

The diagramming of the city in terms of sociospatial neighboring relations raises important questions in regard to spatial practice and urban mobility. Let us take the case of social location, which commonly promotes its services as tools for urban exploration. Dennis Crowley, Foursquare's CEO, explains the remediation of the city through these platforms in this way:

> People are giving us one or two or three pieces of data everyday about the places they go to. We can cut that data up. This is a new way to look at your neighborhood based on the places you've been, and your friends have been to. Places that people like you go to. I can look at the East Village of New York in an entirely different way because the Foursquare algorithm redefined the city for me.[24]

This "redefinition" of the city through personalization, that is, the sorting of urban space according to our previous patterns of mobility and those of our nearest neighbors, although on the one hand might open up new possibilities of "spatial browsing" and curation, on the other hand this might well lead to an intensification of a homophilic experience of the urban encounter. In this respect, Vicsek and colleagues shed light on how the NN rule may influence the spatial behavior of individual agents and its overall influence on the behavior of a networked system.[25] The researchers devised a simulation model of mobile autonomous agents, all moving with similar speed and in random directions upon which the NN rule is applied to control the heading of each agent. The simulation shows that "the nearest neighbor rule . . . can cause all agents to eventually move in the same direction despite the absence of centralized coordination."[26] In this way, when the NN rule considers a distributed network of autonomous agents, it introduces a vector of centralization, which can be interpreted in terms of homogeneity because all agents end up driven toward a single direction. Put in Lefebvre's words, "abstract space is not homogeneous; it simply has homogeneity as its goal, its orientation, its 'lens.'"[27] We could speculate that, in this way, urban mobility could be circumscribed mostly to clusters of neighboring entities.

Navigating the world through clusters of affinities—which is particularly the case of social location platforms where the social navigation of space is a system's default and, to a lesser extent, that of local search—is problematic in terms of the diversity and richness of information and recommendations about places we get. Research on recommender systems has identified a homogeneity problem inherent to these systems, the so-called diversity-accuracy dilemma: the more accurate the results, the less diverse they are.[28] In this regard, homogeneity has also been studied under the term "filter bubble" to refer to the effects the trend toward the personalization of our communications has on the type of information we get, allegedly reinforcing already constituted identities and narrowing worldviews.[29]

The cluster diagram's inherent communication dynamics can be examined more closely by looking at the modeling of epidemics on networks. In this regard, Mark Newman studied the dynamics of epidemic spreading in clustered-network topologies vis-à-vis random network topologies. The model implemented shows that

> in clustered networks epidemics will reach most of the people who are reachable even for transmissibilities that are only slightly above the epidemic threshold. This behavior stands in sharp contrast to the behavior of ordinary fully mixed epidemic models, or models on random graphs without clustering, for which epidemic size shows no such saturation. It arises precisely because of the many redundant paths between individuals introduced by the clustering in the network, which provide many routes for transmission of the disease, making it likely that most individuals who can catch the disease will encounter it by one route or another . . . however, the many redundant paths between vertices when clustering is high make it easier for the disease to spread, not harder, and so lower the position of the threshold. Thus clustering has both bad and good sides where the spread of disease is concerned. On the one hand clustering lowers the epidemic threshold for a disease and also allows the disease to saturate the population at quite low values of the transmissibility, but on the other hand the total number of people infected is decreased.[30]

If we extrapolate these results to an analysis of information transmission, it can be argued that in clustered-network topologies, information spreads easily and rapidly to cluster members (near neighbors), whereas the chances of the same information spreading to other clusters decrease, meaning that a homophilic tendency modulates communication dynamics in the cluster diagram. Studies of social location software usage have already pointed out these homophilic tendencies in mediated spatial practices[31] and have raised concerns about the potential impacts of homophily on the richness of urban life.[32] Although there is a cultural assumption that homophily is positive, which may allegedly be the case in some software applications (e.g., social networking and online dating), from an ethicopolitical perspective, the facilitation of diverse encounters in public spaces remains a desirable democratic ideal of sociability. Even though the cluster diagram produces order and continuity in a disordered and discontinuous urban environment by establishing neighborhoods of similarity, it is, however, the encounter with disordered and diverse urban environments (composed of peoples from different social and ethnic backgrounds), the American sociologist Richard Sennett argues, that best contributes to fostering not only a richer experience of the city but the development of a more mature citizenship. Here Sennett sees "the promise of disorder" in that "in extricating the city from preplanned control men will become . . . more aware of each other."[33] Hence the importance of design practices and tactical interventions that integrate (encode) or explore the potential for diversity and intersubjectivity in the systems mediating our urban experience.

Abstract space, Lefebvre argues, "transports and maintains specific social relations."[34] In the cluster diagram—through the operation of forced clustering—homophilic-oriented relations are fostered and maintained while xenophilic-oriented relations are discouraged. The homophily principle, whereby similarity among nodes fosters network connections, is not specific to locative networks, however. From information transmission to friendship, homophily is pervasive across many different types of social networks.[35] The argument put forward here is rather that inasmuch as it translates geographical distance into social distance, the NN rule represents the computational embodiment of homophily embedded in our new cardinality systems, guaranteeing a homogeneous continuity of sociospatial relations. From this point of view, what is problematic is not that there is homophily in locative networks but that the NN rule constitutes a form of enforced homophily (i.e., an algorithmically homogenized social space)—in the sense that, though accepted as an access condition to these systems, it remains non-negotiable—on the way we sort and navigate the world.

Scott Lash draws attention to algorithmic rules in his account of new forms of posthegemonic power. He uses the term "generative rules" to distinguish them from other types of regulative rules. To Lash, even though these rules are virtual, they have the power to generate actuals.[36] This poses the question of whether this topological continuity might actually translate into a topography through the spatial practices of location-based services users. The possibility that "actions predicated on the basis of topologically abstracted patterns of data" (e.g., ways of perceiving and navigating the urban environment) end up "enacted into existence."[37] If we were to follow the diagrammatic logic delineated here, we might expect to see the spatialization of forms of life and, with it, the deepening fragmentation of social space into homogeneous clusters and hence social stratification. In view of that, locative media users are more likely to be presented, therefore, with a phenomenological experience of urban environments that, as Thrift suggests, is "much closer to a staged performance in which to perceive the environment is also to perceive oneself."[38]

CONCLUSION

Nonetheless, and despite the insights provided by these models, it is crucial to stress the fact that the cluster diagram is but an abstract and deterministic model of locative networking. The diagram disguises a user's agency as well as the "noise" constituent of all human communication. It also does not account for the serendipity of spatial practices, which even some location-based services try to introduce into the programming of their systems. In this sense, the diagram proscribes incompatibility and hence contestation. The cluster diagram might not even be an accurate representation of the actual configuration of sociospatial relations as mediated in locative media because in the diagram this configuration is simply reduced to its ideal form. Moreover, it can be argued that the diagram might make us lose sight of the particularities of its different embodiments, the varied socio-cultural contexts where it is embedded, and the phenomenologically rich and even alternative contesting media practices of users. Even so, I claim, a topological analysis remains important for the diagram as a shaping tendency materially inscribed in these systems and, as such, needs to be problematized in terms of its implied cultural politics. That is to say, topological analysis permits the interrogation of the moment when computational logics may translate into cultural logics, or vice versa.

NOTES

1. Rob Kitchin and Martin Dodge, *Code/Space: Software and Everyday Life* (Cambridge MA: MIT Press, 2011).
2. I borrow the notion of social space from Massey, who understands social space in terms of "the articulation of social relations which necessarily have a spatial form in their interactions with one another." Doreen Massey, *Space, Place and Gender* (Minneapolis: University of Minnesota Press, 1994), 120.
3. Roger Burrows and Nicholas Gane, "Geodemographics, Software and Class," *Sociology* 40, no. 5 (2006): 808.
4. Carlos Barreneche, "Governing the Geocoded World: Environmentality and the Politics of Location Platforms," *Convergence: The International Journal of Research into New Media Technologies* 18, no. 3 (2012).
5. Anastasios Noulas, Salvatore Scellato, Cecilia Mascolo, and Massimiliano Pontil, "Exploiting Semantic Annotations for Clustering Geographic Areas and Users in Location-Based Social Networks," *The Social Mobile Web* 11 (2011).
6. Noulas, Scellato, Mascolo, and Pontil, "Exploiting Semantic Annotations," 33.
7. Noulas, Scellato, Mascolo, and Pontil, "Exploiting Semantic Annotations."
8. Jon Goss, "Marketing the New Marketing: The 'Strategic' Discourse of Advertising for Geodemographic Information Systems," in *Representations in an Electronic Age: Geography, GIS and Democracy*, ed. John Pickles (New York: Guildford Press, 1995), 148.
9. Mark Monmonier, *Spying with Maps: Surveillance Technologies and the Future of Privacy* (Chicago: University of Chicago Press, 2004), 146–147.
10. Gilles Deleuze, *Foucault* (Minneapolis: University of Minnesota Press, 1988), 34.
11. Deleuze, *Foucault*, 36–37.
12. As cited in Richard Harris, Peter Sleight, and Richard Webber, *Geodemographics, GIS, and Neighborhood Targeting* (Chichester, UK: Wiley, 2005), 161.
13. Harris, Sleight, and Webber, *Geodemographics*. See also Emma Uprichard, Roger Burrows, and Simon Parker, "Geodemographic Code and the Production of Space," *Environment and Planning A* 41, no. 12 (2009): 2823–2835.
14. Harris, Sleight, and Webber, *Geodemographics*, 161–162.
15. Alexander D. Singleton and Paul A. Longley, "Geodemographics, Visualisation, and Social Networks in Applied Geography," *Applied Geography* 29, no. 3 (2009).
16. Thomas Cover and Peter Hart, "Nearest Neighbor Pattern Classification," *IEEE Transactions on Information Theory*, 13, no. 1 (1967): 22.
17. Bernard W. Silverman and M. Christopher Jones, "E. Fix and J. L. Hodges (1951): An Important Contribution to Nonparametric Discriminant Analysis and Density Estimation: Commentary on Fix and Hodges (1951)," *International Statistical Review/Revue Internationale de Statistique* 57, no. 3 (1989): 233–238.
18. See Waldo R. Tobler, "A Computer Movie Simulating Urban Growth in the Detroit Region," *Economic Geography* 46, (1970): 234–240.

19. See Carlos Barreneche, "The Order of Places: Code, Ontology and Visibility in Locative Media," *Computational Culture* 2 (2012), accessed September 12, 2013, http://computationalculture.net/article/order_of_places.
20. Annemarie Mol and John Law, "Regions, Networks and Fluids: Anaemia and Social Topology," *Social Studies of Science* 24, no. 4 (1994): 643.
21. Google's patent for its local search algorithm reads: "[a geographical] entity with an elevated density of neighboring entities [annotations] has a greater value than would otherwise be the case" (US Patent 7933897).
22. See Foursquare Engineering, "Machine Learning with Large Networks of People and Places," March 23, 2012, accessed July 22, 2012, http://engineering.foursquare.com/2012/03/23/machine-learning-with-large-networks-of-people-and-places/.
23. Matthew Fuller and Andrew Goffey, "Digital Infrastructures and the Machinery of Topological Abstraction," *Theory, Culture & Society* 29, no. 4–5 (2012): 311–333.
24. Gigaom, "Foursquare's Dennis Crowley: Location Will Connect Us," December 27, 2010, accessed June 7, 2013, http://gigaom.com/2010/12/27/how-location-will-define-ourdigital-experiences-interview-with-foursquare-co-founder-dennis-crowley/.
25. As cited in Ali Jadbabaie, Jie Lin, and A. Stephen Morse, "Coordination of Groups of Mobile Autonomous Agents Using Nearest Neighbor Rules," *IEEE Transactions on Automatic Control*, 48, no. 6 (2003): 988.
26. Jadbabaie, Lin, and Morse, "Coordination of Groups," 988.
27. Henri Lefebvre, *The Production of Space*, trans. Donald Nicholson-Smith (Oxford: Blackwell, 1991), 287.
28. Tao Zhou, Zoltán Kuscsik, Jian-Guo Liu, Matúš Medo, Joseph Rushton Wakeling, and Yi-Cheng Zhang, "Solving the Apparent Diversity-Accuracy Dilemma of Recommender Systems," *Proceedings of the National Academy of Sciences* 107, no. 10 (2010): 4511–4515.
29. See Eli Pariser, *The Filter Bubble: What the Internet Is Hiding from You* (New York: Penguin, 2011).
30. Mark E. J. Newman, "Properties of Highly Clustered Networks," *Physical Review E* 68, no. 2 (2003): 5.
31. Lee Humphreys, "Mobile Social Networks and Urban Public Space," *New Media & Society* 12, no. 5 (2010): 763–778.
32. Alice Crawford, "Taking Social Software to the Streets: Mobile Cocooning and the (An-)Erotic City," *Journal of Urban Technology* 15, no. 3 (2008): 79–97.
33. Richard Sennett, *The Uses of Disorder* (New York/London: Norton, 1970): 198.
34. Lefebvre, *The Production of Space*, 50.
35. See Miller McPherson, Lynn Smith-Lovin, and James M. Cook, "Birds of a Feather: Homophily in Social Networks," *Annual Review of Sociology* 27, no. 1 (2001): 415–444.
36. Scott Lash, "Power After Hegemony Cultural Studies in Mutation?" *Theory, Culture & Society* 24, no. 3 (2007): 71.
37. Matthew Fuller and Andrew Goffey, "Digital Infrastructures and the Machinery of Topological Abstraction," *Theory, Culture & Society* 29, no. 4–5 (2012): 327.
38. Nigel Thrift, *Non-Representational Theory: Space, Politics, Affect* (London: Routledge, 2007): 94.

Part 3

Information, Privacy, Policy

9 Evolving Concepts of Personal Privacy

Locative Media in Online Mobile Spaces

Timothy Dwyer

INTRODUCTION

Our concepts of privacy change constantly in response to their technological and therefore their social contexts. The valorization of privacy is thus an historical construct shaped through the evolution of media technologies, usage forms, and practices. In this second decade of the twenty-first century, the sweeping power of national governments to legislate and to take unilateral privacy-invasive measures on a grand scale in the guise of homeland security has emerged as emblematic of a big data surveillance and privacy zeitgeist. New locative media privacy is inevitably shaped by the contested platform politics of vested political, economic-technological, and social interests.

In this chapter, I argue that, with the rise of web-based media, social networking, and the rapid take-up of mobile devices and apps, notions of privacy are being modified at a commensurate rate for media audiences. It is important to realize that powerful market-dominating new media corporations, such as Google (the owner of YouTube), Facebook, LinkedIn, and Twitter, have made it clear that it is their avowed intention to reconfigure people's understanding of the meanings of personal privacy. This usually incremental change process can be witnessed in continuous website and software updates by these corporations, in the developments in handset design and operations, and in the changing ways that people privately use media devices on the move in public spaces.[1] New mobile media, when viewed as assemblages of hardware, software, and usage practices, are actively implicated in a process of redefining the social and cultural meanings of the concepts we generically label as concerning "privacy." The extension of digital media affordances to gather, to stockpile, and to track and monitor people's usage data in online mobile spaces is pushing out our understandings of privacy into uncharted directions. Many of the privacy concerns that arise for locative media overlap with those of online mobile media by dint of their common transmission infrastructures and access devices. The growth of location-based services (LBS) has been linked with the rise in smartphone ownership and in people getting location-based directions and information for purchasing goods or services, or using them to check in on social media applications while they are on the move.[2]

However, conceptions of privacy linked to the use of location-aware services, apps, and mobile devices can be considered as a key policy issue arising from the broader social and cultural implications of "networked locality."[3] These developments in privacy and the use of personal information are highly consequential as populations increasingly conduct their lives in and through online mobile media transaction spaces, for entertainment, information, and services, for banking and shopping, and for social interactions. Arising from these developments, policy makers need to be alert to the shifting categories of mediatized practices involving personal information and be prepared to specifically identify and corral these for priority interventions. As people depend more and more on global positioning systems (GPS) to "pull" and have information "pushed" to their geolocation, the risks to personal privacy arising from these practices will only increase. Community research in Australia indicates that the majority of people have only a poor level of awareness of the way in which their personal data is shared when they use LBS. Yet this greater use of LBS "does not equate to a greater understanding" about what personal data is collected and shared and with whom, how it was collected, where data is sent, stored or compiled, or indeed who is in control of their personal data.[4] Before we consider more specific privacy concerns raised by LBS, it is important that we have an understanding of the industrial and social contexts in which these new technologies and their associated cultural forms are embedded. It is also important to reflect on the broader ideas regarding privacy itself, as well as the impacts of promotional or selling cultures.

PRIVACY IN CONVERGENT MEDIA INDUSTRIES

Convergent media industries are ushering in new media privacy practices, including those linked with the affordances offered by locative media, and these are merging and diffusing across media platforms together with the transmedia audiences using new screen devices and mobile interfaces.[5] In online mobile media spaces, we are witnessing paradigmatic industrial convergence across telecommunications, radiocommunications, broadcasting, computing, and publishing, as well as in specific service industries such as finance, retail, and other payment sectors, social networking, and entertainment functions, potentially all on a single device.

Changes in privacy are evolving along with the new media and communications industries themselves and with their specific promotional cultures.[6] It is no secret that the major new media corporations, Google and Facebook, have a poor track record on privacy (e.g., Google's Street View or Wi-Fi scandals and Facebook's ongoing "frictionless sharing" tweaks), and they are by no means alone. Media audiences are experiencing these shifts in the contracted and negotiated practices of personal information in searching (e.g., Google, Yahoo, Amazon, eBay, Bing), in social networking (e.g.,

Facebook, Twitter, LinkedIn, Pinterest), in video sharing (e.g., YouTube, Vimeo), in image sharing and archiving (e.g., Flickr, Instagram), and, inseparably, in locative geoweb social media applications (e.g., Facebook Places and Facebook Poke, Foursquare, Google+, Google Maps, Loopt). Instagram and Twitter have a geosocial check-in capability using the Foursquare API. These overlapping industry sectors and associated practices have become the primary new online mobile media spaces for contesting ideas of mediatized privacy. The standard operating procedure for these social media websites is to offer a so-called privacy policy, in part as a requirement of privacy laws and partially as a voluntary consumer-friendly practice (and the application of consumer law). These privacy policies are often then built into the website's terms of service agreements that bind users through either a "browsewrap" or a "clickwrap" consent process.[7] Typically, these contracts contain fine-print references to the categories of sharing and multiple uses of personal information, including by third parties.

Papacharissi argues that the convergent media tend to provide affordances with varying degrees of agency for individuals in the social spaces of available media.[8] She claims that by using interactive and mediated services, the private sphere intersects with public sphere activities. In this way, individuals may, for example, "simulate the private domesticity of the home" in other physical locations, including in the online mobile mediated sphere. But whether this agency extends to control over the flow of all personal information, including in locative media contexts where interactions and data flows depend on terms of service contracts, is, of course, debatable.

That the activities of the child audience are the first major focus of regulation should come as no surprise. As is often the case when it comes to new media and their regulation on behalf of the child audience, social game apps have become the latest site of intense regulatory gaze by the United States Federal Trade Commission (FTC), and they have also sought to strengthen the online privacy rights of children (e.g., in antitracking measures contained in the Children's Online Privacy Protection Act [COPPA]). Similarly, Australia's peak consumer advocacy body, the Australian Communications and Consumer Action Network (ACCAN), has submitted a complaint to the main competition and consumer regulator, the Australian Competition and Consumer Commission (ACCC), regarding the way that in-app purchasing has emerged in game apps for children.[9] For good reasons, the vulnerability of the child audience is an enduring issue for media regulation, irrespective of access device, operating system, or software.

As with digital content cloud management systems, locative media apps get to "know" their users, their likes, dislikes, and detailed consumption patterns. With the passage of time, many of these so-called intelligent interactions occur while people are on the move, and devices such as the iPhone routinely collect, use, and disclose personal information to third parties. Commercialized media storage, distribution, and information and entertainment systems amass and rely on more personal consumer data for additional business purposes and further revenue growth. In the process, third-party businesses and applications will often appropriate personal information and use it in ways that people are unaware of and therefore without having given their consent.

This is an important area for consumer education and regulation alike. The challenge will be to inform mobile app users that, beyond their direct functional uses, for communications, information, entertainment, and productivity, apps have a capacity to amass constant and detailed information about consumers and mobile online activities[10] (see Figure 9.1[11]).

Changing conceptions of privacy should prompt us to ask what is it about privacy that continues to make it so important as a valued moral right? Nissenbaum argues that it is possible to distinguish two general approaches to its worth. The first is an understanding that finds privacy playing a central role in supporting other moral and political rights. The

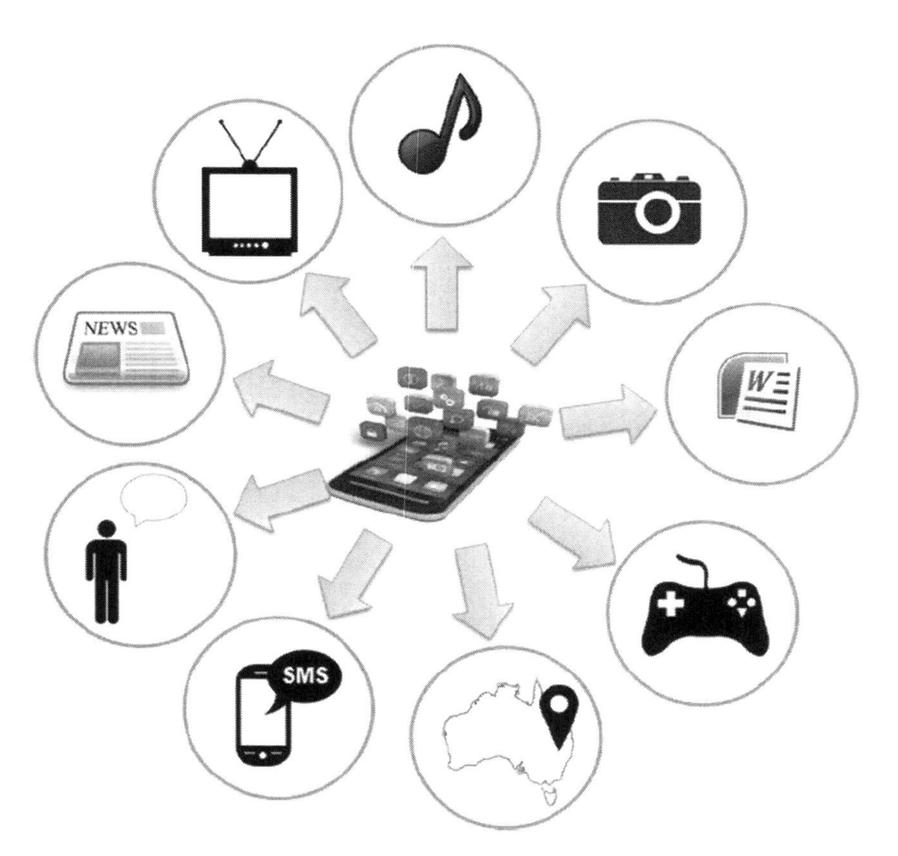

Figure 9.1 Examples of mobile applications functionality.

second "locates privacy's value in the critical role it plays protecting the sphere of the private."[12] Her approach to privacy is guided by the claim that the notion of "contextual integrity" explicates our understanding of privacy. That is to say, people's reaction to how their personal information is appropriated is closely related to the context and purpose of the disclosure and new media and communications technologies themselves. Further, for her this means that "finely calibrated systems of social norms, or rules, govern the flow of personal information in distinct social contexts (e.g. education, health care, and politics)." Her argument is based on the context relativity of informational norms and on how these structure key relationships, at times protecting people from the harmful abuses of power. They are relative because they are "responsive to historical, cultural, and geographical contingencies" that vary among societies. However, it can be argued that her understanding of the context relativity of privacy has its own limitations because it has little to say about the control of the flows of information. Like much of the privacy debate, it tends to be framed in individualistic and proprietorial ideas of data rather than in collectivist or societal conceptions.

Nonetheless, it is clear that the values we attribute to "privacy," or in some contexts "data protection," underpin and often support other long-standing moral, political, and human rights and provide the conditions on which the private and public spheres are sustained. These latter conditions have been argued by some privacy theorists to shore up a much broader set of collectivist ideas of privacy as a public good. Priscilla Regan, for example, argues that individualistic notions of privacy should be replaced with arguments concerning the broader benefits for society as a whole. She has in mind the ways in which privacy supports democratic political systems through the facilitation of anonymous speech and freedom of association. Moreover, privacy can shield people from the intrusive agents of government because it allows people to distinguish their interests, hobbies, tastes, and so on in their private lives from the interests they share in common with fellow citizens.[13]

To reconnect this to our thinking about privacy values in locative media, from both the citizen and the consumer perspectives, we need to ask both broader questions in relation to emerging media and communications services. How has privacy been reconstructed through the terms and conditions of service agreements and other contracts, policies, and codes of practice? And we should ask other, more granular questions, such as what are some of the long-term privacy consequences of massive-scale databases that, by using social media and apps, we have assisted in creating while on the move? And what risks are associated with our data flowing to undisclosed third parties? It is clear, for instance, that higher privacy risks arise when identifiable personal information gets combined with geolocational data, which then may be accessed by sharing networks such as Facebook Places, Foursquare, and other such platforms.

PROMOTIONAL CULTURE AND PRIVACY

A significant proportion of the innovation we are witnessing in new media can be seen to be linked to what McChesney, Stole, Foster, and Holleman refer to as the "sales effort."[14] By this they mean a number of interrelated social, economic, and political components of the wider marketing-media industrial complex that are shaping influences on relations with audiences. On one level, these authors are referring to the ubiquity of advertising forms of content, but on another level it is an intrinsic part of and underpins the commercial media system. But its affects are even more consequential. "If anything, the sales effort is ever more desperate to imprint itself on our brains, and ethical standards are in an uphill battle for survival."[15]

It is into this context that emergent forms of new media advertising and of targeted or behavioral advertising and search need to be situated. As I have previously argued:

> [i]t is clear that the rise of Web 2.0 mechanisms, new "hyper-targeting" technologies using "ad-serving platforms", constitutes a turn in the evolution of commercial speech of media corporations. They are the new media way of corralling and engaging with audiences in the "mass self-communication" contexts of "conversational" Internet social media.[16]

Major components of the media industries, of course, are the advertising industries. Traditional advertising mechanisms are being reconfigured in the digital era as media organizations seek new methods of extracting revenue from their audiences.[17] The advertising revenues that media corporations have depended on for more than a century are declining rapidly, and advertisers are developing new ways to spend their money in the digital era. These newer forms of advertising work by targeting and tracking the consumption habits of specific individuals. This restructuring of media industries is highly consequential for normative assessments of personal privacy.

Digital advertising is a prime mover behind how our ideas of privacy are being modified. The digital ad market is growing far more rapidly than the rest of the advertising market. Total digital advertising (including mobile) rose to US$37.3 billion in 2012 in the United States, a 17 percent increase. Mobile advertising is a rapidly growing sector within this broader digital advertising market. The Pew State of News Media reports that in the U.S. market, for example, mobile ads grew 80 percent in 2012, to approximately US$2.6 billion, and this is expected to grow exponentially.[18]

The five largest Internet media companies—Google, Yahoo, Facebook, Microsoft, and AOL—dominate online advertising and absorbed 64 percent of all digital ad spending in the United States in 2012. As the authors of the Pew report note:

> The digital giants, particularly Facebook and Google, collect and mine vast amounts of data on their users' hobbies, interests, demographic profiles, political interests and relationships. Every time a user "likes" a post on Facebook, conducts a search on Google, or watches a video on YouTube (which Google owns), the companies gain additional data they can use to identify their users' interests . . . both companies offer advertisers a far more sophisticated ability to target specific ads toward consumers than most media companies can muster—and they can increasingly do so in real time by tracking users as they surf the Web.[19]

Not surprisingly, Google has quickly become the dominant player, taking in 54.5 percent of all mobile ad spending in the United States. While Facebook lags behind Google—it introduced its first mobile-only ad feature in June 2012, allowing advertisers to buy ads just on mobile devices—it is growing rapidly. Mobile ads accounted for almost a quarter of Facebook's US$1.33 billion in ad revenues in the fourth quarter of 2012, up from virtually nothing early in the year.[20]

Already advertisers using Google and others can take advantage of the location-based data embedded in mobile devices to place ads targeted to where mobile users are and to what they are doing. Search now brings around half of all mobile ads (about double the next category, banner ads), up from about a quarter as recently as 2009.[21] Its success is all about ad targeting, and it is very successful at delivering very specific audiences to advertisers. As tech firms and others rapidly expand their ability to track consumers as they move around the web or access information on mobile devices, significant new privacy concerns have also arisen.

MOBILES, PRIVACY STAKEHOLDERS, AND DO NOT TRACK (DNT)

Mobile media involve multiple industry stakeholders for whom privacy will have several interconnected but divergent implications depending on their specific roles and use of software for personally identifiable data collection.

First, platform or operating system providers are able to allow app developers access to significant quantities of personal data that emanate from mobile device usage through their application programming interface (API). Second, apps stores, such as Windows Store, Apple's App Store, Google Play, or the Android Market, play an important role in communicating and making available privacy policies to users. Third, app developers have a key point-of-contact responsibility to provide disclosures regarding data and express consent. App developers need to ensure that the communications they have with third-party service providers such as ad networks and ad data brokers is clear, so that they in turn can provide accurate information to app consumers. Fourth, the wider advertising networks and other third parties, such as trade associations and data analytics providers, also need to have good channels of communication so that information remains accurate (see Figure 9.2[22]). Inasmuch as mobile carriers provide the infrastructure, they therefore have key consumer educational responsibilities in relation to personal data flows.

The advisory note, "This Site Uses Cookies," is familiar to most Internet users. With the rise of the Internet, the use of cookies has always been an important component in privacy debates. Now with the growth of Web 2.0 and mobile advertising, cookies are again under scrutiny by privacy regulators and advocacy organizations. So-called behavioral advertising is a particularly powerful tool for the advertising industry when personalized media devices are the mode of access. For instance, online usage statistics now indicate that more people are accessing Facebook on mobile devices than on laptops or other fixed-use computers. The Australian Communications and Media Authority (ACMA), in a recent research paper, notes that, "within the first 15 minutes of waking up, four out of five smartphone owners are checking their phones. Of these, almost 80 per cent reach for their phone before doing anything else."[23]

But how do cookies actually harm individual privacy? It is worth revisiting how cookies work on the Net. As Facebook's privacy policy explains: "Cookies are small files that are placed on your browser or device by the website or app you're using or ad you're viewing."[24] Facebook's procookie corporate spin notes that, "Like most websites, we use cookies to provide

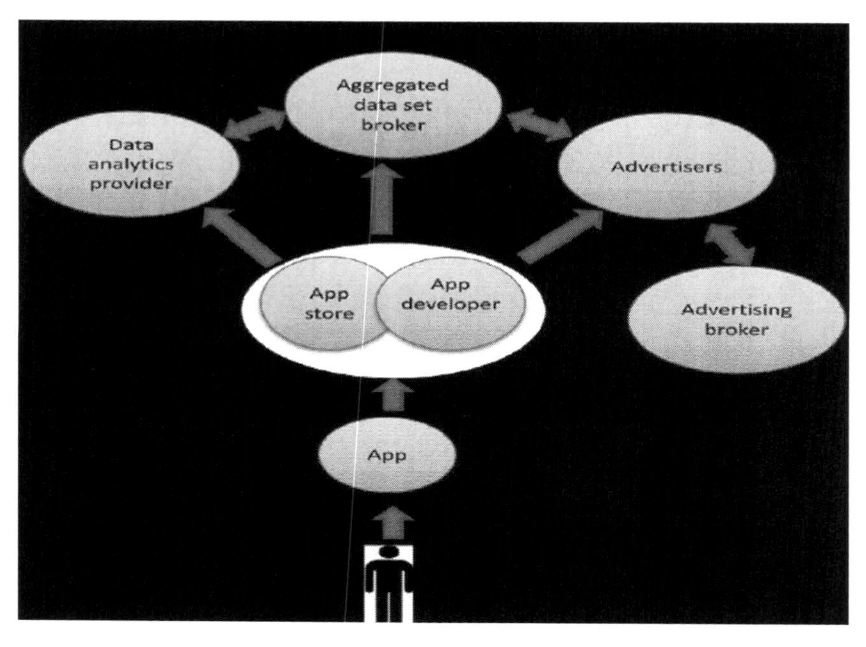

Figure 9.2 Mobile application information flows.

you with a better, faster or safer experience."[25] In plain language, cookies enable online advertising by collecting information about the pages we visit, the ads we click on, the products we select for our virtual supermarket trolleys, or simply the interest we show by clicking on and reading their blurbs. All this activity enables companies to build a profile of our interests and preferences and then serve tailored ads to us. For John Keane, this indicates that, in an age of monitory democracy, "users of the Internet find their personal data is the engine fuel of a booming web-based market economy."[26] Indeed, personal data has been likened to the "new oil" of this economy where "[t]he collection and analysis of anonymised location and behavioural information to develop user profiles and personalised marketing material is broadening the meaning attached to personal information."[27]

Regulatory responses vary from one jurisdiction to another and from self-regulation to stronger forms of regulation in this emerging area of online mobile and locative media. In Ireland, for example, their Advertising Standards Authority introduced new rules in late 2013 under which companies are required to make it clear they are gathering information about individuals' online activities and to show an icon indicating that they are doing so.[28] In the United States, the Federal Trade Commission (FTC) has urged the industry to adopt stronger do-not-track policies that would allow consumers to prohibit web sites, data firms, and advertising technology companies from collecting and mining personal data based on their online activity. These rules are similar to earlier rules developed for online behavioral advertising targeting children, which is now banned under the updated COPPA. Express parental notice and consent are required for this kind of advertising or marketing, including retargeting ads based on browsing history. The expanded definition of "personal information" under these laws includes geolocation data, images, and any video or audio content.[29]

The FTC also recommended that the mobile industry build similar safeguards into the software and apps used on smartphones and tablets. For now, however, users have only limited ability to control how their personal data is collected or used. These developments are all within the realm of emerging do-not-track (DNT) privacy policies, which require platforms, app developers, and service providers to take consumer choices about targeted mobile ads into account. The FTC is actively working with mobile industry stakeholders to further develop DNT options. For example, Apple's iOS now allows consumers to use a "limit ad tracking setting."[30] Specifically, in relation to online mobile media, it has requested that platform or operating system providers "consider offering a DNT mechanism for smartphone users."[31] The report authors argue that this would "allow consumers to choose to prevent tracking by ad networks or their third parties as they navigate among apps on their phones."[32]

Yet in spite of growing awareness, several of the major Web 2.0 corporations continue to attract regulatory scrutiny. A high-profile case in point is Google, which has long been the focus of competition and data protection regulators' investigations around the world. Google routinely collects a wide variety of personal information, including names, images, e-mail addresses, phone numbers, credit cards, websites visited, device used, search engine queries, phone numbers called, as well as their time, date, duration, and location.

Now questions of privacy and the use of personal information by Google are a central focus by European privacy agencies, including the Agencia Espanola de Proteccion de Datos in Spain, France's Commission Nationale de l'informatique et des Libertes (CNIL), and the UK's Information Commissioner's Office. While Spain has charged the search giant with up to €1.5 million in fines for data privacy breaches, the French have ordered it to rewrite its privacy policies under threat of fines of over €300,000. They also want "definite retention periods," limits on data being combined from different products and services, and for users to be informed about the use of cookies to track browsing. It is reported that 37 data protection agencies, including Canada, Australia, Israel, Switzerland, and New Zealand, were all signatories to a combined letter to Google requesting that the company build privacy considerations into the development of products and services, including the new Google Glass product, which exhibits locative affordances.[33] The capacity for modifying privacy norms should not be underestimated from the company that popularized Google Earth, or "a gift to terrorism" as it has also been referred to.[34]

LOCATION-BASED SERVICES

Local digital advertising is a growth sector in these changes, and geotargeting of audiences is now an important component of advertising strategies.[35] The Pew researchers argue that "to maintain a strong share of local digital ads, news organizations will need to increase their own ability to target or work with the ad networks."[36]

LBS are a growth subsector that holds great promise for revenue in the advertising sector.[37] It has moved from a fringe practice on the Internet involving art or gaming to mainstream activities, including Sat Nav devices in cars, to Google Maps on desktops and mobile phones, to geotagging and the geoweb, and to iPhone and Android apps that use location-based apps. Geotagging has become a key part of being on social media, and its popularity appears to be about equal between genders and fairly evenly spread across age groups[38]. However, in their report, the Pew Center noted that a slightly greater number of 30- to 49-year-olds tag location in social media posts. Perhaps more significantly, 16 percent of teens set their profiles to tag automatically, while almost a half of all teen app downloaders have turned off the location-tracking features.[39]

People's data and their interactions with the locations they move through using locative media are becoming important for the media-marketing industrial complex. One example of the myriad of burgeoning LBS entertainment services available is the hugely popular and now profitable Shazam, the music, TV, and ad-identifying app that is fun to use in various recreational contexts, such as pubs or clubs. An innovative and globally expanding UK-based "media engagement" company owns the app that is reported to have been used by 350 million people around the world, with 70 million active monthly users and that picks up new users at the rate of 2 million each week.[40] It is an example of an entertainment LBS app that is very well suited to a hybrid traditional–new advertising revenue model, where the media company delivers vast audiences to advertisers. The aural samples look up databases for a match, and the sound bites can also work in a similar way to a swiped QR code to deliver tailored campaign deals and messages.

Yet these commercial location-based services (LBS) can have their share of attendant risks: geolocation features in particular have important privacy implications for providers of news apps. Wilken notes the stark example of the traffic from Foursquare reappearing on Please Rob Me.com, a site that aggregates data to highlight the risks of posting location information[41] (see also Chapter 12, this volume). The FTC has been researching and developing policies for mobile phone privacy for several years. In a report released early in 2013, *Mobile Privacy Disclosures: Building Trust Through Transparency,* the agency has made a series of best practice recommendations for industry actors. The recommendations focus on the practices of platform/operating system providers, application developers, advertising networks, data analytics companies and other third parties, and trade associations in relation to effectively making privacy disclosures and obtaining express consent.

In early 2012, the attorney general for California announced an agreement that committed leading operators of mobile application platforms to improve privacy protections for millions of consumers who use apps on their smartphones, tablets, and other mobile devices.[42] The agreement—entered into with the six companies that dominate the majority of the mobile apps market: Amazon, Apple, Google, Hewlett-Packard, Microsoft, and Research In Motion—was to ensure that these companies with their mobile apps have privacy principles in a privacy policy that is easily locatable on their websites and that comply with the California Online Privacy Protection Act.

It is apparent that news apps (and other LBS) using location data to push out specific consumer information or advertising in categories such as business opportunities, real estate, or travel, will raise privacy risks for individuals. As the FTC *Mobile Privacy Disclosures* report notes:

> Mobile devices can reveal precise information about a user's location that can be used to build detailed profiles of consumer movement over time and in ways not anticipated by consumers. Indeed, companies can

> use a mobile phone device to collect data over time and reveal the habits and patterns that mark a day in the life and a way of life. Even if a company does not intend to use the data in this way, if data falls in the wrong hands, the data can be misused and subject consumers to harms such as stalking or identity theft.[43]

Similarly, a joint policy document issued by Canadian Information and Privacy Commissioners noted:

> Whatever method is used to link a device to its owner, whether it's a unique device identifier or multiple linked identifiers, it has the potential to combine with personal information to create a profoundly detailed and sensitive profile of a user's behaviour depending on the circumstances.[44]

So there is a growing expectation that the developers of mobile apps will need to inform users of apps with geolocation features of what information about them is being collected, used, and disclosed. These policy processes call for transparency and openness and for meaningful consent in the use of personal information. Various practices are already in place that signal to users that their geolocation data is being used by apps. For example, at the platform level, both Apple and Google currently use icons to communicate to users when an app is accessing their geolocation information.[45]

CONTEXTUAL USE STRATEGIES FOR CONNECTED CULTURES

The shifting definitions, understandings, and priorities for governance of privacy discussed in this chapter infer that an equal level of ambiguity will inevitably prevail in privacy debates. You might argue that this suits those major corporations whose economic lifeblood is personal information. On the other hand, many legal instruments have been enacted at a supranational and national level, and specific rights and legal obligations that flow from these justify the ongoing updating of privacy frameworks that address the harms wrought by mediatizing technologies and practices. If the terms-of-service contracts available in online mobile media generally, and locative media spaces in particular, offer few or limited unenforceable remedies, then this will require further education in the first instance about practical options, as well as more specific reform interventions.

A continuous stream of major policy reports recommending specific legislative frameworks (or a tort of privacy in those jurisdictions that do not already have these laws) represent a growing tide of opposition to the encroaching moves of the information data miners. We are witnessing a battle for the control over the extent to which commercial media corporations in these new media contexts control our personal information where habituated people willingly barter their data lives away. Our understandings of privacy in online mobile media spaces are being stretched in new ways that complement sociotechnical, cultural, and economic change. For policy makers, the long-term strategies needed to address morphing conceptions of privacy and rights will need to include developing a catalog of specific contextual uses, where harm to personal privacy is likely to accelerate. Recent responses seen in the U.S. FTC and the Canadian OPC reports, as well as the actions taken by regulators around the world, signpost the first steps along this path. But, in the absence of these kinds of policy strategies, the dilution of existing privacy expectations and norms in mediatized contexts is inevitable. There is simply too much political economic power pushing locative media applications for smartphones, tablets, and other mobile media devices for it to be otherwise.

Immersed as we are in the so-called cultures of connectivity of online mobile spaces, we need to promote the idea that people can tactically conceive of their interactions as mutually constitutive of those practices and platforms.[46] Yet in some measure, Bauman's "liquid modernity" is an apt metaphor for a postprivacy society, where there's a kind of privatized ambivalence and widespread acceptance of a transition in our own self-mediatizing and communicative practices. Locative media is undoubtedly a catalyst for building the foundations for new privacy normativities. On the other hand, to accept this logic uncritically is also to reject the long history of hard-won productive privacy reforms, where the harms to people, caused through the disclosure of personal information, have been ameliorated by laws, policies, and regulatory frameworks.

NOTES

1. Adriana De Souza e Silva and Jordan Frith [in *Mobile Interfaces in Public Spaces: Locational Privacy, Control, and Urban Sociability* (New York: Routledge, 2012)] analyze the growth of locative media as interfaces and explore the implications of these changes for our understanding of personal privacy.
2. Kathryn Zickuhr, "Location-Based Services," *Pew Research Center* (September 12, 2013), accessed October 11, 2013, http://pewinternet.org/Reports/2013/Location.aspx.
3. Adriana de Souza e Silva and Eric Gordon, *Net Locality: Why Location Matters in a Networked World* (London and New York: Routledge, 2011).
4. ACMA, *Locational Services, Personal Information and Identity: Exploratory Community Research* (Sydney: ACMA, 2012), 3.
5. Tim Dwyer, "Net Worth: Popular Social Networks as Colossal Marketing Machines," in *The Propaganda Society: Promotional Culture and Politics in a Global Context*, ed. Gerald Sussman (New York: Peter Lang, 2011), 77–92.
6. Browsewrap consent arises from additional browsing past a homepage, whether or not the user is aware of their implicit agreement to terms or not. Clickwrap agreements involve an active click of assent to the terms of service and related policies. Woodrow Hartzog, "Privacy and Terms of Use," in *Social Media and the Law. A Guidebook for Communication Students and Professionals*, ed. Daxon R. Stewart (New York: Routledge, 2013), 56.

7. Zizi Papacharissi, *A Private Sphere: Democracy in a Digital Age* (New York: Routledge, 2010), 63.
8. ACMA, *Mobile Apps: Emerging Issues in Media and Communications* (Sydney: Commonwealth of Australia, 2013), 8, fn. 16.
9. "An app which purports to perform a limited range of basic functions for a user may also access a range of other data on the consumer's device that is unrelated to the 'primary' function of the app." Figure 1 in ACMA, *Mobile Apps*, 6.
10. Figure 1 in ACMA, *Mobile Apps*, 6.
11. Helen Nissenbaum, *Privacy in Context: Technology, Policy and the Integrity of Social Life* (Stanford, CA: Stanford University Press, 2010), 13.
12. These arguments of Priscilla Regan are from her book *Legislating Privacy* (Chapel Hill, NC: University of North Carolina Press, 1995), cited by Nissenbaum, *Privacy in Context*, 86. They are inspired by the work of Hannah Arendt, John Rawls, Oscar Gandy, and Carl Friedrichs.
13. Robert W. McChesney, Inger L. Stole, John Bellamy Foster, and Hannah Holleman, "Advertising and the Genius of Commercial Propaganda," in *The Propaganda Society: Promotional Culture and Politics in Global Context*, ed. Gerald Sussman (New York: Peter Lang, 2011), 28.
14. McChesney, Stole, Foster, and Holleman, "Advertising," 38.
15. Tim Dwyer, "Net Worth: Popular Social Networks as Colossal Marketing Machines," in *The Propaganda Society: Promotional Culture and Politics in Global Context*, ed. Gerald Sussman (New York: Peter Lang, 2011), 78.
16. See Joseph Turow, *The Daily You: How the New Advertising Industry Is Defining Your Identity and Your Worth* (London: Yale University Press, 2013).
17. Jane Sasseen, Kenny Olmstead, and Amy Mitchell, "Digital: As Mobile Grows Rapidly, the Pressures on News Intensify," *Pew Research Center* (2013), accessed October 11, 2013, http://stateofthemedia.org/2013/digital-as-mobile-grows-rapidly-the-pressures-on-news-intensify. There are similar growth figures for mobile advertising in the UK: "Mobile advertising is forecast to almost double in 2013 to 1bn pounds, as Google, Facebook and Twitter increasingly successfully mine the smartphone, tablet and app revolution." Mark Sweeney, "UK Mobile Advertising Market Set to Grow By 90% in 2013," *The Guardian* (June 24, 2013), accessed October 11, 2013, www.guardian.co.uk/media/2013/jun/24/uk-mobile-advertising-market-double. YouTube is reported to have tripled its mobile ad growth in the same period as a result of its app being removed from the constraints of the iOS software platform. Stuart Dredge, "YouTube's Mobile Advertising Takes Off," *The Guardian* (June 6, 2013), accessed October 11, 2013, www.guardian.co.uk/technology/appsblog/2013/jun/06/youtube-mobile-advertising.
18. Sasseen, Olmstead, and Mitchell, "Digital."
19. Sweeney, "UK Mobile Advertising."
20. Dredge, "YouTube's Mobile Advertising."
21. Figure 8 in ACMA, *Mobile Apps*, 19.
22. The following report was cited by ACMA: International Data Corporation, *Always Connected: How Smartphones and Social Keep Us Engaged* (2013), accessed October 11, 2013, https://fb-public.box.com/s/3iq5x6uwnqtq7ki4q8wk.
23. Facebook, *Data Use Policy: Cookies, Pixels and Other Similar Technologies* (2013), accessed October 11, 2013, www.facebook.com/about/privacy/cookies.
24. Facebook, *Data Use Policy.*
25. John Keane, "Power and Privacy in an Age of Monitory Democracy," *The Conversation* (July 4, 2013), accessed October 11, 2013, http://theconversation.com/power-and-privacy-in-the-age-of-monitory-democracy-14695.

26. ACMA, *Privacy and Personal Data: Emerging Issues in Media and Communications* (Sydney: Commonwealth of Australia, 2013), 1.
27. Pamela Newenham, "Firms to Give Notice If Collecting Online Data for Ads," *The Irish Times* (June 24, 2013), 2.
28. Bryron Acohido, "The Ripple Effects of Stricter Privacy Rules for Kids," *USA Today* (July 1, 2013), accessed October 11, 2013, www.usatoday.com/story/cybertruth/2013/07/01/coppa-rules-children-tracking-behavioral-targeting-ftc-child-safety/2479815/.
29. Federal Trade Commission, *Mobile Privacy Disclosures: Building Trust Through Transparency* (February 2013), 20, accessed 11 October 2013, www.ftc.gov/reports/mobile-privacy-disclosures-building-trust-through-transparency-federal-trade-commission.
30. Federal Trade Commission, *Mobile Privacy Disclosures*, 20.
31. Federal Trade Commission, *Mobile Privacy Disclosures*, 20.
32. Google, *How It Feels Through Glass* (2013), accessed 11 October 2013, www.google.com/glass/start/how-it-feels/.
33. Evgeny Morozov, *The Net Delusion: How Not to Liberate the World* (London: Penguin, 2011), 237.
34. Sasseen, Olmstead, and Mitchell, "Digital."
35. Sasseen, Olmstead, and Mitchell, "Digital."
36. See Adriana de Souza e Silva and Eric Gordon, *Net Locality: Why Location Matters in a Networked World* (Chichester, UK: Wiley-Blackwell, 2011); Rowan Wilken, "Locative Media: From Specialised Preoccupation to Mainstream Fascination," *Convergence: The International Journal of Research into New Media Technologies* 18, no. 3 (2012): 243–247, doi: 10.1177/1354856512444375.
37. Sasseen, Olmstead, and Mitchell, "Digital."
38. Zickuhr, "Location-Based Services," 14–16.
39. Stuart Dredge, "Shazam Raises £26.9m for Latin America Expansion with Carlos Slim," *The Guardian* (July 9, 2013), accessed October 11, 2013, www.guardian.co.uk/technology/appsblog/2013/jul/08/shazam-funding-america-movil.
40. Wilken, "Locative Media," 243.
41. State of California Department of Justice: Office of the Attorney General, *Attorney General Kamala D. Harris Notifies Mobile App Developers of Non-Compliance with California Privacy Law* (October 30, 2012), accessed October 11, 2013, http://oag.ca.gov/news/press-releases/attorney-general-kamala-d-harris-notifies-mobile-app-developers-non-compliance.
42. Federal Trade Commission, *Mobile Privacy Disclosures*, 20.
43. Office of the Privacy Commissioner of Canada, *Seizing Opportunity: Good Privacy Practices for Developing Mobile Apps* (October 2012), accessed 11 October 2013, www.priv.gc.ca/information/pub/gd_app_201210_e.asp.
44. Federal Trade Commission, *Mobile Privacy Disclosures*, 17.
45. José Van Dijck, *The Culture of Connectivity: A Critical History of Social Media* (New York: Oxford University Press, 2013).

10 Google Glass and Australian Privacy Law

Regulating the Future of Locative Media

James Meese

INTRODUCTION

Google Glass (Glass)—a wearable computer with a heads-up display (HUD)—has raised significant privacy concerns following Google's soft launch of the project within a small group of early adopters, engineers, and journalists. The hardware, which allows users to take photos and videos or live-stream to the Internet through the glasses' interface, has already been banned from a number of cafes, casinos, bars, and movie theaters.[1] Footage of an arrest in progress was also captured through Google Glass, sparking further debate about the ramifications of this new technology.[2] Unsurprisingly, the publicity has caught the attention of regulators across the world, and Google has had to formally respond to questions from government and data protection authorities about the impact of this new technology.

The introduction of Google Glass reveals clear points of tension among social norms, existing regulatory frameworks, and the role of companies like Google within this environment. In this chapter, I examine these tensions by examining the impact of Google Glass generally, before going on to discuss a specific regulatory and policy context in which the hardware may operate. I will begin by outlining the current privacy concerns over Glass and the emergence of locative media on the Glass hardware. I will then outline Google's interactions with the wider policy discussion about privacy and the international policy response to Glass, before outlining the current state of privacy law in Australia. The chapter will end by considering the impact of a possible statutory right to privacy in Australian law on technologies like Google Glass. My analysis will provide a detailed account of how a jurisdiction has attempted to develop a regulatory response to the emergence of locative media, as well as considering how commercial entities like Google might navigate these changes when attempting to introduce new location-centric hardware like Glass.

LOCATIVE MEDIA, PRIVACY, AND GOOGLE GLASS

Locative media applications have started to appear on Google Glass, disclosing one's location to service providers or peers in a similar manner to existing location-based applications.[3] Google Glass is not GPS-enabled yet, with users having to tether their Glass to a compatible Android smartphone in order to access GPS and enable accurate mapping. However, developers are already using this configuration to develop locative media through Glass. The most notable example is Field Trip, a "geo-publishing tool" that brings up relevant information about places nearby in the manner of a "deconstructed guide book."[4] It has already received significant attention from media, for whom the app represents the potential future of augmented reality. There are also persistent rumors that Google is planning to introduce GPS technology to Glass in the near future, directly tying a users' location to the use of the hardware.[5] Considering the potential success of a location-based social media application and the recent release of the Glass development kit to third-party developers, which now allows developers to "tap into real-time location," it is clear that location will play a central role in future iterations of Glass.

What is most interesting about Glass in the context of location is that it represents a significant advance as a form of locative media. Carlos Barreneche notes the general "movement towards a geocoded word," with "spatial annotation" allowing media to constitute "a form of metadata about the world."[6] However, unlike earlier platforms, such as the smartphone where this type of annotation is seen as a feature, spatial annotation is integral to the operation of Glass. One's location can be linked to the broader environment seamlessly when using Glass and through applications like Field Trip subsequently mobilized in order to enhance basic practices of spatial navigation like walking through a museum or in a park. Furthermore, it appears that future iterations of Glass will also attempt to account for a user's embodied, as well as spatial, location. Google has recently filed a patent that would potentially allow Glass to track eye-movements, monitor emotions, and the types of advertisements that users look at.[7] This patent represents a further shift from earlier forms of locative media. Glass incorporates elements of a users' corporeal body into the broader political economy around locative media, introducing embodied notions of location and attention into the commercial sphere.

Glass therefore stands as a significant step forward in Google's attempt to commercialize location and to intervene in the new search markets that are emerging around mobile technologies and location.[8] Google is heavily reliant on advertising revenue, and both mobile searches and location-aware technologies are becoming increasingly important tools to advertisers. Google's own research has acknowledged this fact, noting that, following a search query, "nine out of ten mobile search users take some action, with over half leading to an act of consumption."[9] Location plays a key role in this process, and with locative search preeminent in the "mobile media experience," local advertisers look to convert this "data traffic into foot traffic."[10] When considering the privacy of Glass users when negotiating this new technology, one must keep this wider political economy of location in mind. Users are not necessarily disenfranchised, having given up their spatial and embodied location willingly. Yet it is worthwhile noting that when Glass is used, through Google's monopolized economy of "place," a commercial relationship is automatically formed among a user's data, Google, and the wider commercial marketplace.[11]

As well as affecting Glass users' locational privacy, the Glass interface also reshapes nonusers' perceptions of spatial locational privacy. Much of the public furor over the technology has centered on the innocuous appearance of the glasses and the difficulty of being able to tell whether someone is filming or taking a picture.[12] Therefore, the glasses also raise questions about the asymmetrical power relations that emerge when a Google Glass wearer interacts directly with the rest of the population, who are unable to participate in an equal form of reciprocal surveillance.[13] The willingness of business owners to ban the glasses from their premises underlines this power imbalance, with one of the first owners to ban Glass (the owner of a dive bar) noting that "we don't let people film other people or take photos unwanted of people in the bar, because it is kind of a private place that people go."[14]

The "kind of" qualifying the statement of the bar owner is interesting because it underlines the situated and contextual nature of privacy, which is "not about zeroes and ones, [but] about how people experience their relationship with others and with information."[15] As existing work has outlined, questions of privacy are centered on issues of control and context rather than on a strict delineation between public and private knowledge.[16] This conceptual framework helps us understand the initial public reaction to Glass. Despite recent revelations that uncovered the true extent of government surveillance on the Internet,[17] much of the furor over Google Glass has centered on the problem of "collateral surveillance," that is, surveillance conducted horizontally by people rather than institutions.[18] The public reaction to Glass is understandable once we consider the small number of Glass owners. The clear imbalance between Glass owners and the rest of the population makes the lack of "control" over one's personal privacy much more pronounced.

GOOGLE AND PRIVACY: FROM STREET VIEW TO GLASS

These privacy issues, which have emerged as Glass has started to be used more widely, carry a particular resonance when viewed in the context of Google's recent engagements with privacy regulation. The gradual introduction of Google Street View led to similar debates on how to protect privacy in the face of perceived technological overreach. The Street View project saw Google cars, equipped with rooftop cameras, taking photos of streets in particular countries in order to offer users "a 360-degree street-level panorama of a given point of a map, including whatever and whomever happened to be caught on camera at the instant the photos were taken."[19] The project received a varied international response. European countries objected strongly, and protests were staged in the United Kingdom, Switzerland, the Czech Republic, and Greece, with authorities called on to investigate the practice.[20] Germany responded in the strongest fashion, with the objections of their state and federal politicians temporarily halting the development of Street View across Europe.[21] Subsequently, Google gave advance warning of the locations it intended to photograph across Europe,[22] and limited the introduction of Street View in Germany to 20 cities.[23] Conversely, the response in the United States was muted, with authorities placing few restrictions on the photography of American homes and citizens from the street.

However, what was initially a questionable project became a scandal when it was revealed that as well as photographing streets, Google cars were also recording data from private, unencrypted Wi-Fi networks as they photographed. Google publicly acknowledged their mistake in 2010, and their announcement subsequently led to a series of independent investigations conducted by various national privacy authorities. It is worth noting that the punishment meted out to Google varied widely, signaling vastly different cultural values assigned to notions of privacy. Google received no fine from Australian regulators,[24] and the U.S. Federal Communications Commission could find no violation of American law.[25] Conversely, Johannes Caspar—the data protection supervisor of Hamburg, Germany—undertook a dogged three-year investigation of Google and fined the company US$189,225.[26] Considering this problematic history, it was no surprise that privacy authorities viewed the introduction of Glass skeptically and that an immediate regulatory response followed.

Members of the U.S. congressional privacy caucus sent a letter to Google in May 2013, asking for further information about the privacy implications of the new technology,[27] noting Google's ongoing problems with privacy protection. Following the recent issues with Google Street View, the letter questioned how information would be secured in Google Glass.[28] The caucus also inquired about the possibility of facial recognition technology on Glass and suggested that Google may wish to amend their privacy policy considering the capabilities of Glass, among other things.[29] In responding, Susan Molinari (Google vice president, public policy and government relations) noted that "protecting the security and security of [their] users [was] one of [their] first priorities."[30] She also explained that Google would not be approving any facial recognition Glassware at this time and that the company expected social norms to eventually develop around Glass as they did for "cell phones, laptops [and] cameras."[31] No changes to the Google privacy policy were planned for Glass at that time, according to Molinari.

A group of data protection authorities—including the Privacy Commissioner of Australia—conducted a similar regulatory intervention soon after the congressional caucus. In a letter to Google, they noted that the group was not targeting Google but that, due to their leadership in the area, Google happened to be the first company to "confront the ethical issues that such a product entails."[32] They asked a similar set of questions, inquiring about Glass's compliance with data protection laws, the privacy safeguards that were currently in place and asked Google to address the "broader social and ethical issues raised by such a product."[33] The response from Google again was clear, that they would develop the technology first and assume regulatory and social norms would be adapted as Glass continued to be used.

Google's response to these attempted interventions sheds some light on its complex position as a commercial entity within this regulatory space. Google aims to be perceived as responsible corporate citizen, a goal evident in its own informal motto "don't be evil." The company regularly engages in public policy development and articulates these goals strongly when discussing questions of Internet freedom and the rights of the online individual. Google's stance on these issues was most notably aired in its response to the recent National Security Agency, with Google stating that they were "outraged at the lengths to which the government seems to have gone to intercept data from our private fiber networks."[34] However, as these exchanges with regulatory bodies outline, the company takes a vastly different approach when it comes to privacy issues that impact on its own commercial interests. Google's strategic approach to the question of privacy becomes particularly interesting as we now turn to an assessment of these regulatory debates in the Australian context. Australia stands as a unique case study because there is no common law protection of privacy, no constitutional protection for either privacy or free speech, as might be found in the United States,[35] and privacy rights are rarely enforced.[36] This complex legal arena therefore provides a useful space from which to assess the potential introduction of Glass and to place Google's own rhetoric about privacy regulation in a specific local context.

GOOGLE GLASS AND PRIVACY LAW IN AUSTRALIA

To first provide a brief outline of privacy law in Australia, it is important to note that privacy is protected by "a patchwork of specific legislation"[37] that currently offers little protection to the Australian public as they attempt to negotiate the increased use of locative media. At the federal level, the Privacy Act 1988 (Cth) covers both the public and private sector, respectively. In addition, legislation in five of the seven states and territories—New South Wales, Victoria, the Northern Territory, Tasmania, and Western Australia—offers enforceable privacy rights in the public sector. However, there is a lack of "authoritative interpretation" by either the relevant courts or state and federal privacy commissioners. This is because of the narrow framing of Australia privacy law, which means that cases concerning privacy are often "catalogued and argued as cases about . . . some other legal head."[38] With no foundational principles on which to develop a fundamental right of privacy, interpreting key provisions of many of the acts is an entirely speculative exercise.[39]

This lack of a conceptual framework for privacy means that currently only limited redress is available for members of the Australian public if they feel their privacy has been infringed. This point is worthwhile reflecting on as we consider the emergence of new platforms for locative media such as Google Glass. Currently, any breach of privacy through the use of these technologies would be addressed by the Federal Privacy Commissioner. This authority regulates the handling of information by large private sector organizations such as Google, as well as "any business that collects or discloses personal information for a benefit, service or advantage," which would account for businesses using personal information like one's location in order to sell advertising.[40] However, the function of the Privacy Commissioner makes it particularly difficult for the public to assert their privacy rights.

Strangely enough, if users of Google Glass felt that their privacy was being infringed through the misuse of data, the first port of call when making a complaint about a privacy breach would be either Google or the application developer that had breached their privacy. This is because, if a direct resolution has not been attempted by the individual initially, the commissioner may refuse to investigate further. Furthermore, if the commissioner feels that the business or government agency has adequately dealt with the issue, the case can be closed, regardless of how the complainant feels about the issue.[41] A complainant is able to appeal to the Federal Court of Australia or the Federal Magistrates Court; however, the Court will not review the merits of the case but merely decide whether the matter should be referred back to the commissioner.

This process attempts to close complaints before moving to the legal arena and is useful in the sense of providing a rapid response to complaints, but it also means that the law is rarely tested and that formal decisions are rarely made. In practice, the majority of cases regarding the private sector are "closed without investigation" and that compensatory damages are rarely awarded, which suggests to businesses that they have "little to fear" from the Commissioner.[42] The process therefore provides a friendly operating environment for companies such as Google, while not offering adequate protection for members of the public. Australia has alternative legal mechanisms to protect one's privacy, such as breach of confidence. However, this approach is not a transparent and obvious way for the general public to navigate and protect their privacy rights. Its appropriateness as a legal remedy has also been questioned.[43]

In recent years, Australian legislators have responded to these ongoing problems with privacy law, and, in mid-2008, the Australian Law Reform Commission conducted the first major review of privacy law in 20 years. Following significant public consultation, the center-left Labor Government enacted the Privacy Amendment (Enhancing Privacy Protection) Bill

2012. The bill, which went into effect March 12, 2014, allows a clear right of appeal against the Commissioner's decisions and permits the Commissioner's office to begin its own investigations and require Privacy Impact Assessments (although, these assessments do not need to be made public). However, Nigel Waters and Graham Greenleaf note that, historically, Commissioners have rarely used enforcement powers that they have been given and that the Commissioner still retains unwarranted discretion to refuse to investigate a complaint.[44] Despite the intention of these reforms, the centrality of the Privacy Commissioner in enforcement-related matters means that currently the Australian public are unable to enforce their privacy rights in the most effective manner.

CHANGING CULTURES: A STATUTORY TORT VERSUS GOOGLE GLASS?

Australian legislators have continued to engage with these ongoing problems with privacy law, and a current Australian Law Reform Commission (ALRC) inquiry has been tasked with assessing the impact of locative media. "Serious Invasions of Privacy in the Digital Era," the supporting issues paper,[45] notes that existing issues considered by various law reform commissions, such as the "unauthorised use of personal information and [the] intrusion on personal privacy and seclusion," must now be reexamined in light of

> [n]ew technologies [that] allow unprecedented levels or surveillance and tracking of the individuals, of recording and communication of personal information, and of intrusion into personal space.

The ALRC raises the possibility of implementing a statutory cause of action for invasion of privacy, but what would this mean for Google Glass or indeed for locative media more generally?

The introduction of a statutory tort would potentially result in the increased litigation of privacy breaches. In contrast to Australia's current privacy law, with its weak enforcement methods, a statutory tort would present a clear legal remedy for claimants. The particular uses of Glass could be impacted by these reforms, but this would largely depend on what law considered a "serious" a breach of privacy to be. In *ABC vs. Lenah Game Meats*, Gleeson CJ noted that certain kinds of information "relating to health, personal relationship or finances, may be easy to identify as private" and similarly "many kinds of activity . . . a reasonable person . . . would understand [are] to be unobserved." In addition to this notion of "reasonableness" outlined in legal doctrine, the statutory tort could also test "seriousness" or "offensiveness," discouraging litigation over minor matters.[46] These proposed standards raise a number of interesting questions concerning privacy law, and these additional tests would also have a significant impact on a statutory tort.

The major problem with this approach is that it would present an "onerous and unfair standard for a privacy claimant" and "carve out a protected zone of privacy-intrusive free speech."[47] As Megan Richardson notes, rather than assessing the breach on its merits, the majority will would be "the proper standard of what a privacy expectation should be," although Richardson concedes that this language of "offensiveness" could be interpreted differently in Australia.[48] Regardless, offensiveness sets a very high standard for a breach of privacy action. Indeed, such an approach would ignore usage of Glass, which may simply be considered intrusive or unwelcome. Making a statutory tort actionable *per se* would potentially lead to more cases but would carry the added benefit of developing a legal discourse about privacy, which in Australia has been sorely lacking to date. Considering the uniqueness of Glass, which could see new challenges emerge beyond the existing locational privacy issues that currently define locative media, a statutory tort stands as an opportunity for a body of law to develop alongside social norms at this nascent stage.

Google[49] has responded to the ALRC issues paper and suggested that if a statutory tort were enacted, it should be available only:

- To natural persons;
- In circumstances where the person has a reasonable expectation of privacy;
- Where the act is sufficient to cause substantial offense to a person of ordinary sensibilities; and,
- Where the act complained of was intentional or reckless.

This willingness to consider a statutory tort is a positive step, but it is notable that Google also suggests that a defense of consent be introduced. Although Google notes that they have an online privacy center "which explains in plain English what data Google stores and how we use it," the notion that a user knowingly consents to the use of their private data is a problematic one.[50] More often than not, users "agree to privacy policies which are nearly incomprehensible," and while contractually they offer their assent, they often do not know exactly how their location data is being shared.[51]

Therefore, transparency and clarity become important issues to consider in this space. The ALRC commissioner of the inquiry, Barbara McDonald, acknowledges that a statutory tort will not stop breaches of privacy wholesale, so the general public must be able to navigate these corporate privacy policies relatively easily.[52] However, as the ALRC notes, the collection of location data, in particular, has been poorly understood by consumers in the Australian context. The Australian Communications and Media Authority recently reported that consumers who are using location services regularly were less likely to display an understanding of the range of options for protecting information, and almost half the respondents (46 percent) "were not aware of the data-sharing processes that result from accessing location services."[53] This lack of knowledge carries a direct impact on a potential statutory tort of privacy because, by consenting to an unwieldy privacy policy, Glass users could preclude an actionable claim.

These findings underline the fact that many regular uses of locative media lack basic education about the potential privacy implications of this media practice. Google's public statements about the ability of the general public to effectively protect their privacy and agree to the sharing of their data also appear as misguided in the context of this research. As Daniel Solove notes, if this form of privacy self-governance is going to be the main way that consumers can protect their privacy, law may need to play a more active and educative role in guiding users through these policies, especially in regard to "new uses [of data] at the time that they are proposed."[54] This type of soft intervention would be incredibly useful for users of Glass and locative media users more broadly. By taking an educative as well as regulatory role, regulators would be able to ensure a greater understanding of the impact of locative media among users, as well as maximize the effectiveness of the proposed tort.

CONCLUSION

Australia does not have Europe's strong regulatory culture concerning privacy, nor does it rest on a constitutional foundation, as in the United States.[55] Subsequently, the legislative framework that has developed around privacy has been comparatively narrow and conceptually weak. The potential introduction of a statutory right therefore stands as an interesting example of how a jurisdiction traditionally averse to strong privacy regulation is responding to the possible futures of locative media. Hardware such as Glass outlines a radical future of location, where location is tied to one's corporeal and phenomenological existence. It has also already instigated changes in how we conceive of privacy at particular locations, with the possibility of live-streaming or taking photos through glasses presenting a challenge to existing social norms. This emergent regulatory culture concerning privacy in Australia indicates that, in particular jurisdictions, it is simply not enough for social norms concerning locative media to change. Instead, the existing legal culture in Australia also requires new regulatory strategies in order to provide an adequate response to potential harms.

Location has become an important part of everyday media practice in a relatively short time. Companies like Google have attempted to balance their own corporate self-interest with a general commitment to protecting legal rights online. However, as noted throughout this article, the commercial environment on its own does not provide sufficient protection or guidance for an increasingly location-centric public. Locative media also present a series of problems for jurisdictions like Australia, which, up to this point, has not developed a sufficient legal discourse about privacy. Therefore, in closing, I suggest that Australia should introduce a statutory tort alongside a broader program of legal education on locative media. These reforms will offer sufficient redress for the wider Australian public which, in a time of rapid change, currently do not have a legal framework robust enough to anticipate these possible futures of locative media.

NOTES

1. Jessica Guynn, "Google Glass Sees All and That's a Worry," *Los Angeles Times* (August 11, 2013), accessed December 11, 2013, http://articles.latimes.com/2013/aug/11/business/la-fi-google-glass-20130811.
2. Connor Simpson, "What the First Arrest Captured on Google Glass Really Means," *The Wire* (July 7, 2013), accessed December 11, 2013, www.thewire.com/technology/2013/07/what-first-arrest-captured-google-glass-really-means/66901/.
3. See Eric Gordon and Adriana de Souza e Silva, *Net Locality: Why Location Matters in a Networked World* (Chichester, UK: Wiley-Blackwell, 2011).
4. Alexis C. Madrigal, "The World Is Not Enough: Google and the Future of Augmented Reality," *The Atlantic* (October 25, 2012), accessed December 11, 2013, www.theatlantic.com/technology/archive/2012/10/the-world-is-not-enough-google-and-the-future-of-augmented-reality/264059/.
5. Zach Honig, "Google Glass to Support GPS Navigation, Text Messages Without Companion App," *engadget* (May 3, 2013), accessed December 11, 2013, www.engadget.com/2013/05/03/google-glass-gps-text-messages/.
6. Carlos Barreneche, "Governing the Geocoded World: Environmentality and the Politics of Location Platforms," *Convergence: The International Journal of Research into New Media Technologies* 18, no. 3 (2012): 332.
7. Steve Hawkes, "Google in Privacy Row over Glasses That Can Monitor Eye Movements." *The Telegraph* (August 15, 2013), accessed December 12, 2013, www.telegraph.co.uk/technology/google/10245396/Google-in-privacy-row-over-glasses-that-can-monitor-eye-movements.html.
8. See Rip Empson, "WTF Is Waze and Why Did Google Just Pay a Billion+ for It?" *TechCrunch* (June 11, 2013), accessed December 11, 2013, http://techcrunch.com/2013/06/11/behind-the-maps-whats-in-a-waze-and-why-did-google-just-pay-a-billion-for-it/.
9. Barreneche, "Governing the Geocoded World," 340.
10. Barreneche, "Governing the Geocoded World," 340.
11. Barreneche, "Governing the Geocoded World."
12. See Heather Kelly, "Google Glass Users Fight Privacy Fears," *CNN* (December 10, 2013), accessed December 11, 2013, http://edition.cnn.com/2013/12/10/tech/mobile/negative-google-glass-reactions/.
13. Jason Farman, *Mobile Interface Theory: Embodied Space and Locative Media* (New York: Routledge, 2012), 70.
14. Casey Newton, "Seattle Dive Bar Becomes First to Ban Google Glass," *CNet* (March 8, 2013), accessed December 11, 2013, http://news.cnet.com/8301–1023_3–57573387–93/seattle-dive-bar-becomes-first-to-ban-google-glass/.
15. danah boyd, "Facebook's Privacy Trainwreck," *Convergence: The International Journal of Research into New Media Technologies* 14, no. 1 (2008): 18.
16. See Gordon and de Souza e Silva, *Net Locality*; Farman, *Mobile Interface Theory*; Adriana de Souza e Silva and Jordan Frith, *Mobile Interfaces in Public Spaces: Locational Privacy, Control, and Urban Sociability* (New York:

Routledge, 2012); Helen Nissenbaum, "Privacy as Contextual Integrity," *Washington Law Review* 79 (2004): 101–158; Helen Nissenbaum, *Privacy in Context: Technology, Policy, and the Integrity of Social Life* (Palo Alto, CA: Stanford University Press, 2009).

17. See "The NSA Files,"*guardian.co.uk,* accessed December 11, 2013, www.theguardian.com/world/the-nsa-files.
18. de Souza e Silva and Frith, *Mobile Interfaces in Public Spaces,* 125.
19. Roger C. Geissler, "Private Eyes Watching You: Google Street View and the Right to an Inviolate Personality," *Hastings Law Journal* 63 (2012): 899.
20. See Geissler, "Private Eyes Watching You"; Jordan E. Segall, "Google Street View: Walking the Line of Privacy-Intrusion upon Seclusion and Publicity Given to Private Facts in the Digital Age," *University of Pittsburgh School of Law Journal of Technology Law and Policy* 10 (2010): 1–32.
21. Geissler, "Private Eyes Watching You," 899.
22. Segall, "Google Street View," 31.
23. David Murphy, "Google Abandons Street View in Germany," *PCmag.com* (April 10, 2011), accessed December 11, 2013, www.pcmag.com/article2/0,2817,2383363,00.asp.
24. "Australian Privacy Commissioner Obtains Privacy Undertakings from Google," Office of the Australian Information Commissioner, accessed December 11, 2013, www.oaic.gov.au/news-and-events/statements/privacy-statements/google-street-view-wi-fi-collection/australian-privacy-commissioner-obtains-privacy-undertakings-from-google.
25. David Streitfeld and Kevin O'Brien, "Google Privacy Inquiries Get Little Co-operation," *New York Times* (May 22, 2012), accessed December 11, 2013, www.nytimes.com/2012/05/23/technology/google-privacy-inquiries-get-little-cooperation.html?pagewanted=all.
26. Claire Cain Miller, "Stern Words, and a Pea-Size Punishment, for Google," *New York Times* (April 22, 2013), accessed December 11, 2013, www.nytimes.com/2013/04/23/business/global/stern-words-and-pea-size-punishment-for-google.html?_r=0.
27. Brendan Sasso. "Lawmakers Fear Privacy Risks from Google Glass," *The Hill* (May 16, 2013), accessed February 10, 2014, http://thehill.com/blogs/hillicon-valley/technology/300303-lawmakers-raise-privacy-concerns-over-google-glass.
28. See Rowan Wilken, "Locative Media: From Specialized Preoccupation to Mainstream Fascination," *Convergence: The International Journal of Research into New Media Technologies* 18, no. 3 (2012): 243.
29. The questions from the bipartisan privacy caucus are quoted in Google's reply. Susan Molinari, letter to Senator Joe Barton (June 7, 2013), accessed February 10, 2014, http://marketingland.com/wp-content/ml-loads/2013/07/Google_Glass_Response_2013_Letter.pdf.
30. Molinari, letter to Senator Joe Barton.
31. Molinari, letter to Senator Joe Barton.
32. "Data Protection Authorities Urge Google to Address Google Glass Concerns," *Office of the Privacy Commissioner of Canada* (June 18, 2013), accessed February 10, 2014, www.priv.gc.ca/media/nr-c/2013/nr-c_130618_e.asp.
33. "Data Protection Authorities Urge Google to Address Google Glass Concerns."
34. Dominic Rushe, Spencer Ackerman, and James Ball, "Reports That NSA Taps into Google and Tahoo Data Hubs Infuriate Tech Giants," *The Guardian*, accessed February 10, 2014, www.theguardian.com/technology/2013/oct/30/google-reports-nsa-secretly-intercepts-data-links.
35. Geissler, "Private Eyes Watching You": 911.
36. See David Lindsay, "An Exploration of the Conceptual Basis of Privacy and the Implications for the Future of Australian Privacy Law," *Melbourne University Law Review* 29 (2005): 131.

37. Graham Greenleaf, "Privacy in Australia," in *Global Privacy Protection: The First Generation*, eds. James B. Rule and Graham Greenleaf (Cheltenham, UK: Edward Elgar, 2008), 148.
38. Michael Kirby, "Privacy—In the Courts," *University of New South Wales Law Review* Forum 7, no. 1 (2001), accessed February 10, 2014, www.austlii.edu.au/au/journals/UNSWLJ/2001/2.html. For a discussion of the doctrine behind alternative legal approaches that Kirby discusses, see also Megan Richardson, "Breach of Confidence, Surreptitiously or Accidentally Obtained Information and Privacy: Theory Versus Law," *Melbourne University Law Review* 19 (1993): 673.
39. Graham Greenleaf, "Privacy in Australia," 152.
40. Graham Greenleaf, "Privacy in Australia."
41. "What Happens to Your Privacy Complaint," *Office of the Australian Information Commissioner*, accessed December 11, 2013, www.oaic.gov.au/privacy/what-happens-to-your-privacy-complaint.
42. Greenleaf, "Privacy in Australia," 162–163. This trend continued in 2012–2013 with the Privacy Commissioner receiving 1,496 complaints, closing 55 percent without investigation and choosing to investigate only 9.4 percent. See Office of the Australian Information Commissioner, *Annual Report 2012–13* (October 31, 2013), accessed February 10, 2014, www.oaic.gov.au/about-us/corporate-information/annual-reports/oaic-annual-report-201213/.
43. David Lindsay, "Playing Possum? Privacy, Freedom of Speech and the Media Following ABC v Lenah Game Meats Pty Ltd.," *Media and Arts Law Review* 7 (2002): 1.
44. Nigel Waters and Graham Greenleaf, "A Critique of Australia's Proposed *Privacy Amendment (Enhancing Privacy Protection) Bill 2012*," University of New South Wales Faculty of Law Research Series (2012), Paper 35: 6.
45. "Serious Invasions of Privacy in the Digital Era Issues Paper," 10.
46. "Serious Invasions of Privacy in the Digital Era Issues Paper," 19. See also Megan Richardson, "When Should Privacy Be Legally Protected?" *Fortnightly Review of IP & Media Law*, accessed December 12, 2013, http://fortnightlyreview.com/2011/11/04/when-should-privacy-be-legally-protected/. Richardson argues that this approach would present an "onerous and unfair standard for a privacy claimant" and "carve out a protected zone of privacy-intrusive free speech."
47. Richardson, "When Should Privacy Be Legally Protected?"
48. Richardson, "When Should Privacy Be Legally Protected?"
49. Iarla Flynn, "Google Australia," submission to Australian Law Reform Commission, "Serious Invasions of Privacy in the Digital Era Issues Paper" (November 25, 2013), 7.
50. Iarla Flynn, "Google Australia": 8.
51. de Souza e Silva and Frith, *Mobile Interfaces in Public Spaces,* 128.
52. Chris Merritt, "Tort Won't Stop Privacy Breaches: Professor," *The Australian* (October 11, 2013), accessed December 12, 2013, www.theaustralian.com.au/business/legal-affairs/tort-wont-stop-privacy-breaches-professor/story-e6frg97x-1226737229468.
53. Australia Media and Communications Authority, *Here, There and Everywhere—Consumer Behaviour and Location Services* (Melbourne: Commonwealth of Australia, 2012), 17.
54. Daniel Solove, "Privacy Self-Management and the Consent Paradox," *Harvard Law Review*, 126, no. 7 (2013): 1902.
55. See Geissler, "Private Eyes Watching You."

11 Locative Media, Privacy, and State Surveillance in Mexico

The Case of the Geolocalization Law

Gerard Goggin and César Albarrán-Torres

> [L]ocative technologies entail not only the potential for new, panoptic ways of seeing on the part of the consumer/subject—who may display, manipulate and author locationally referenced data . . . —but also new ways to *be* coded and made visible. Increasingly, the public is seen and seeing through the digital artifacts of a geocoded life.[1]

Locative media are now widely diffused and increasingly utilized around the world via cellular mobiles (especially smartphones); check-in programs such as Foursquare, promoted outside the United States; geolocative aspects of social media (Twitter and micro blogging, but also Facebook and other social networking programs such as Instagram and Yelp); and increasingly sensors and sensor networks connecting objects that range from smart watches and portable fitness devices to cars.

Yet it could be said that the unfolding of locative media in many important and culturally diverse parts of the world—the Middle East, Africa, the South Pacific, and Latin America, to mention but a few—has been strikingly understudied. Much of the available research on locative media focuses on North America, Europe, and East Asia (especially South Korea and, to a lesser extent, China).[2] Elsewhere, the literature on locative media hinges on a few celebrated cases, such as citizen journalism and news experiments in Kenya (a case in the point is the Ushahidi phenomenon).[3] There is some important work on locative media, especially in relation to locative media arts, coming from Brazil;[4] however, the experience of the rest of South America and Latin America broadly has received little attention. Drawing on accounts of internationalizing media and Internet studies,[5] it could be argued that such gaps in the existing research and theorization constitute a serious obstacle to our understanding of locative media and how they disrupt the social and political understanding of personal data.

To start with, it means that we know little about the forms that locative media are taking in different parts of the world and what the differential patterns of consumption and cultures of use in particular places look like.[6] The political implications of the information generated by locative media in particular national contexts will also elude us.

The knock-on effect of this gap in locative media studies is likely to be that the dominant understanding of locative media rests on the early, celebrated cases and the ideas associated with the inception and take-up of locative media in a narrow range of societies. A genuinely cosmopolitan account of locative media stands to give us a much more accurate map of the technologies and cultures that constitute this phenomenon.

In this chapter, we aim to contribute to internationalizing research on locative media via a critical discussion of a recent Mexican case—the 2011–2014 controversy concerning the *Ley de Geolocalización* (Geolocalization Law). It is important to note that the Geolocalization Law is not a single document, but that it refers to a set of changes to two articles: 133 Quáter of the Federal Code for Penal Proceedings; and Article 16, fraction I, section D, and 40 Bis of the Federal Telecommunications Law.

Geopolitically, Mexico sits at the intersection of two highly developed North American neighbors (the United States and Canada) and the underdeveloped region of Central America. Mexico is situated at the crossroads of diverse migration, linguistic, and cultural flows, making it a handy site for studying local and transnational uses of media.[7] In addition, as we shall explain, the recent crime, security, and political features of the Mexican state and society have combined to put issues concerning personal information and locative media in stark relief.

THE GEOLOCALIZATION LAW: SECURITY VERSUS PRIVACY?

The Mexican House of Representatives passed the Geolocalization Law on September 2011, followed by the Senate in March 2012. The aim of this law was to consolidate and fortify the existing provisions requiring telecommunications providers to make available their phone records and data to police and security agencies for purposes of law enforcement. The law aimed to target drug dealers, kidnappers, and blackmailers who use mobile information and communication technology (ICT) to carry out illegal activities.

In particular, the Geolocalization Law extended existing provisions to require telecommunications carriers to make geolocal data, gathered through cellular mobile devices and networks, available to agencies; hence the title of the law. As we shall see, the Gelocalization Law enjoyed considerable support from lawmakers and some activists in Mexico, a society riven by *el narcoterrorismo*, the parastatal power of the drug cartels that has resulted in horrific, widespread violence, kidnappings (often organized from federal prisons via mobile phones), human rights violations, power voids in states such as Michoacán, corruption, and impunity. However, the Geolocalization Law also drew significant criticism from opponents, who were seriously concerned over the poor track record of the state, law enforcement, and security agencies when it comes to the protection of privacy and human rights. The Geolocalization Law presented lawmakers and activists with a dilemma: should security supersede privacy or vice versa?

This chapter aims not to take sides on the controversial Geolocalization Law but simply to point out how local context determines the ways in which locative media are used by individuals and institutions—not to mention the ensuing negotiations generated by access to new types of information. Such dynamics raise questions. What happens when location *is* content? How does this alter the complex sets of relationships among citizens, media, and the state? How are these social relations concerning locative media unfolding in the Mexican situation? And what lessons do they offer us for directions in locative media, their policy, and research in national and transnational contexts?

MOBILE AND LOCATIVE MEDIA IN MEXICO

Overall, the Mexican mobile media market, on which geolocalization relies, is experiencing a steady expansion, and demand for third- and fourth-generation (3G and 4G) networks is growing. There are four operators in the mobile market: Telcel (owned by Carlos Slim's América Móvil), Nextel, Iusacell, and Movistar (owned by the Spanish company Telefónica). Mobile phones are a ubiquitous technology in Mexico, with 83.35 mobile phone subscribers per 100 inhabitants in 2012.[8] The 3G market comprises approximately 10.2 percent of the mobile market (2013) and is expected to cover 18.1 percent by 2017, amounting to a predicted 92.9 subscribers per 100.[9]

Little research is available on the apps industry in Mexico and, in particular, on the design, take-up, and use of apps with geolocative capacities in Mexico. However, there are a wide range of apps including a significant number that have their parallels elsewhere, such as *Se me antoja* (order home delivery in participating restaurants), *Metro México* (maps of the public transport system in Mexico City, Guadalajara, and Monterrey), and *Guía Pemex*, which allows the user to find the nearest gas station.[10] Geolocative services have also been introduced in Mexico via global platforms, especially Twitter, Facebook, and Foursquare.

What is evident in the introduction of such locative media in Mexico is the prominence of privacy and security concerns and fears in the social imaginary of such technologies. Accompanying the introduction of Foursquare, for instance, some commentators voiced their concerns over the user's personal safety vis-à-vis the high incidence of kidnappings and robberies in the country, which they argued could be facilitated by access to geolocative data.[11] Apps such as *Taxibeat* (http://taxibeat.com.mx), which was originally developed in Greece, respond to these specific concerns in the Mexican context, such as kidnappings in Mexico City taxis. Through Foursquare, *Taxibeat* geolocates users and allows them to call a taxi that is safe and registered. The user can access information such as the driver's name and photograph, the vehicle's real-time location, and user ratings.[12]

In addition to these general uses of locative media, the technology has also featured in law enforcement. The app *Mi Policía* (My Police Officer), launched by Mexico City authorities, helps users locate the nearest cop in case of emergency, whether they are victims, people at risk, or individuals who have just witnessed a crime.[13] *Mi Policía* allows users to communicate with their local police directly and in real time, something that raises questions about the efficacy of police enforcement and notions of the role of citizens in personal and public safety. It has attracted criticism, such as that from Marguerite Cawley from InSight Crime, an activist organization, who are doubtful about this type of use of locative media: "such apps will only benefit a small percentage of the population in Mexico, failing to address violence in poorer, rural areas."[14]

PRIVACY LAW IN MEXICO: *DERECHO A LA PRIVACIDAD*

As we can see, locative media have occasioned distinctive issues in relation to privacy in Mexico. In contemporary Mexico, as much as elsewhere, state access to personal electronic information has incited heated debates in recent times, particularly after the mid-2012 Edward Snowden revelations concerning surveillance by the state agency NSA in the United States. These revelations concerned not only U.S. citizens but also foreign governments and individuals. The Snowden revelations accentuated the rising concerns regarding personal information, particularly after *Der Spiegel* revealed that the NSA hacked into Felipe Calderón's e-mail account in 2010, while he was still the president of México, as well as into the accounts of officials of the Public Security Secretariat in 2009. This gave the NSA access to diplomatic information, as well as data regarding the ongoing conflict involving drug cartels. According to the report, Brazil was also targeted,[15] which led Brazilian President Dilma Rousseff to cancel a state visit to the United States.[16]

Although Mexico has featured in the global concerns about personal information, it needs to be acknowledged that there are special issues in considering privacy in the country. In particular, the general sense of mistrust toward institutions in Mexico has a long, complex history. Highly publicized cases of corruption at the highest levels of local and federal government have taken their toll on the public's trust in officials. There are widely shared concerns—if not a consensus—that organized crime has infiltrated most institutions (especially governmental, regulatory, and financial) and is now a hegemonic power. Some go as far as to dub Mexican politics as "narcopolitics," implying that the cartels have affected governance and stability at the highest levels.

In regard to privacy laws in Mexico and the use of technologies that generate information about and around the individual, Vargas argues that most legal systems "are not yet equipped with principles, legal rules, and institutions especially designed to provide individuals and legal entities with adequate protection against the intrusive and abusive use of these modern technologies."[17] This is particularly salient when discussing the Geolocalization Law.

Vargas points out that, although there is no clear nomenclature related to "privacy rights" (*derecho a la privacidad*) in Mexican law, this right may be included in the set of "individual guarantees" (*garantías individuales*) or "human rights" (the nomenclature used in the Constitution). The Supreme Court has, however, alluded to "privacy rights" (*derecho a la vida privada*) in some cases. This legal blurriness has vast implications in a technosocial context in which "privacy rights appear to be a universal reaction against the unwanted intrusions and serious threats posed by scientific and technological devices that are inimical to certain inherently personal rights".[18]

The Mexican legal framework, however, does offer some indications as to how private information generated by mobile and geolocative media could be protected under current legislation (which constitutes the main argument presented by the National Human Rights Commission against the Geolocalization Law, as we will see). In particular, Article 16 of the Mexican Constitution and a 1996 amendment address the issue of personal information:

> ARTICLE 16. No one shall be disturbed in his or her person, family, domicile, documents or possessions except by virtue of a written order by the competent authority stating the legal grounds and the justification for the action taken.

Two paragraphs were added to Article 16 in 1996, taking into consideration the widespread availability of cellular phones and other forms of networked ICTs, including the following:

> Private communications are inviolable. The law shall criminally sanction any act which infringes upon the liberty and privacy of said communications. Only the federal judicial authority, at the request of the federal authority empowered by the law or the Public Prosecutor [Ministerio Público] of the corresponding federal entity, may authorize the interception of any private communication. For this purpose, the competent authority, in a written order, must provide the legal grounds and the justification of the request, indicating, in addition, the type of interception, the subjects involved, and its duration. The federal judicial authority cannot grant these authorizations in matters of an electoral, fiscal, mercantile, civil, labor or administrative nature, nor in the case of communications between a detainee and his or her legal counsel.

The current Mexican legal framework does not make it clear, however, whether surveillance through geolocalization is considered an invasion of privacy.[19] This background is important for understanding how the Geolocalization Law has emerged and been received.

LA LEY DE GEOLOCALIZACIÓN: LOCATIVE MEDIA AND THE STATE

The Geolocalization Law arose as a response to the unprecedented levels of violence during the war against crime led by president Felipe Calderón in the six years of his government (2006–2012). The law was also a response to the accumulation of long-standing problems with kidnapping and crime against civilians, particularly in Mexico City. Kidnapping and extortion had become common practice among criminal organizations, providing a key source of cartel funding.[20] The Geolocalization Law was promoted by some enduring advocates for victims' rights, such as businessman Alejandro Martí[21] (founder of the advocacy group SOS Ciudadano), and activist Isabel Miranda de Wallace [now a well-known politician under the banner of the conservative Partido Acción Nacional (PAN)]—both of whose children were kidnapped and murdered. Another supporter was the high-profile poet and activist Javier Sicilia, another of the most prominent voices in recent debates on the effects of violence in civil society. The backing of such well-known figures in the struggle against criminal violence gave moral validity to the campaign for the new law.

The Geolocalization Law was proposed in the Senate during Calderón's presidency, on March 15, 2011, by an alliance of senators: Tomás Torres Mercado (a member of the Partido Verde Ecologista de México, a green party associated with the PRI), Fernando Jorge Castro Trenti (from the now incumbent PRI party), and Alejandro González Alcocer (the conservative Partido Acción Nacional, or PAN party). In his speech introducing the legislation, Torres Mercado argued that the economic well-being and potential for employment in Mexico currently could not be divorced from "the theme of penal justice, of the violence of public security."[22] In order that the state would have a suitable legal instrument in delicate matters of kidnapping, extortion, or threats, Torres Mercado contended that those holding a telephone concession and so benefitting from the public property of the nation should be obliged, if possible, "to tell us from where a phone call is being made."[23] In addition, it needs to be possible to disable a phone that has been stolen and also to block telephone calls made from prison for the purpose of extortion. Torres Mercado related the words of Alejandro Martí to the effect that if this initiative was in place "when the kidnapping of my son occurred, perhaps he would still be alive [as] I received a phone call for more than 8 days, every day, consecutively."[24]

Thus, when the Geolocalization Law came into effect over one year later in April 2012, it gave authorities sweeping power to access geolocal data. Authorities were given the right to access geotagging information from any mobile device they suspect was used to commit certain types of crime (including kidnapping, extortion, and organized crime). Service providers, in turn, have the obligation to cooperate with authorities without reservation. Mobile service providers have the obligation to provide enough data to the Federal Attorney's General Office (Procuraduría General de la República,

PGR) to determine the location of people and devices in real time, especially in cases of kidnapping. To achieve these measures, the law amended provisions to the Federal Code of Criminal Procedures, the Federal Criminal Code, and the Federal Telecommunications Law.[25] This action is justified by the Mexican Congress as follows: "[It has the goal of] establishing new tools that provide the Mexican State the possibility of being more efficient in the investigation of felonies related to organised crime, kidnapping, extortion and threats," and "it also tries to discourage the stealing of mobile use and their use for criminal purposes."[26]

The Geolocalization Law also allows authorities to block devices reported as lost or stolen so that they cannot be used by criminals. Devices being used inside prisons will be blocked to prevent their use for purposes of extortion. It is well-known that criminal activities are conducted from penitentiary facilities in Mexico and that mobile telecommunication systems are fundamental in the development and maintenance of criminal networks.

The move to introduce the Geolocalization Law enjoyed considerable high-profile support, especially due to the palpable, widespread sense of the insecurity and vulnerability of everyday life. For instance, according to the National Survey on Victims and Public Perception of Safety, conducted by the National Statistics and Geography Institute (Instituto Nacional de Estadística y Geografía, INEGI) in 2013, 32.4 percent of Mexican households have experienced at least one crime firsthand, 92.1 percent of crimes are not investigated, 72.3 percent of the population feel unsafe, and there are 35,139 crimes per year for every 100,000 citizens. These figures include crimes such as robbery on the street or in public transport, extortion, threats, and fraud. Other illegal activities, such as organized crime, drug dealing, and human trafficking, are not included in this report but are nevertheless taxing in the everyday life of most citizens.

THE ROCKY PATH OF THE GEOLOCALIZATION LAW

As well as notable support for the Geolocalization Law, significant opposition has also emerged. On the face of it, it might sound odd for extensions of police, investigative, and official powers to be opposed, given the parlous state of all sectors of Mexican society given the power, violence, and disruption of the drug cartels. Yet, in reality, such opposition comes as no surprise in a country where trust in the political sphere, historically and present-day, is low and most crimes go unpunished (INEGI, 2013). Initially, the law was described by opinion leaders as well intentioned and necessary. However, the fine print gives the state unprecedented access to private information. This, added to the prevalent and widely shared mistrust in authorities and law enforcement agencies in the country, drew a significant response. Take, for instance, concerns expressed by the blog Contingente MX (http://contingentemx.net/), a platform for the discussion of civil rights issues:

> The real time monitoring of the location of a mobile phone provides a vast array of sensitive information about the user. This is why it is essential that there are sufficient security measures to protect the user's personal data. Without these measures the door is open for actions that far from protecting us from organised crime, increase our vulnerability.[27]

The public debate about the Gelocalization Law was ramped up significantly with the decision of the lead Mexican human rights agency to contest it. In May 2012, the National Human Rights Commission (Comisión Nacional de los Derechos Humanos, CNDH) brought a lawsuit against the law, claiming that it is unconstitutional. The lawsuit, signed by the legal representative Raúl Plasencia Villanueva, details the constitutional and international precepts that, in their view, are being violated. Specifically, CNDH contended that the Geolocalization Law was in breach of the aforementioned Articles 14 and 16 of the Mexican Constitution, Article 11 of the American Human Rights Convention, Article 17 of the International Pact for Civil and Political Rights, Article 12 of the Universal Declaration of Human Rights, and Article 16 of the Convention for Children's Rights. According to CNDH, the Geolocalization Law violated the principles of legality and judicial security, as well as the right to privacy and a private life (CNDH 2012). A further issue for CNDH was that there was no apparent temporal limitation for when the authorities are permitted to access information, something the Commission felt would open the door for constant state surveillance of a subject under investigation (CNDH 2012). Additionally, CNDH suggests that the law lacked clarity about third parties who might be involved in the investigation but not the subject of it—yet who could potentially be surveilled in the very wide casting of the geolocative net.

For CNDH, the Geolocalization Law raises issues that go to the heart of privacy, the private and public spheres, that the role and powers of the state. CNDH notes that the sort of privacy protected by the Mexican Constitution concerns:

> that which is not part of public life, what escapes the action and knowledge of others, what is only shared with whom the subject chooses; every activity conducted in the private sphere, including those related to honor and family or anything that the individual does outside of public service. (CNDH 2012)

Against this right, for CNDH, the Geolocalization Law provides the authority in charge of prosecuting a crime with a tool that by its nature violates the human right to privacy, which constitutes a violation of judicial security (this is not clearly established in the Mexican Constitution, however). In effect, argues CNDH, the law affords an almost limitless access to private information, providing authorities with the right to access a thorough and precise registry of a people's whereabouts and public movements, which reflects "important details about their personal, family, political, religious and social life." CNDH sounds a clarion call, submitting that even though physical intrusion is not necessary considering new technologies such as GPS systems, state surveillance over citizens could alter the relationship between the two in such a way that could be harmful to a democratic society.

Shortly after the CNDH opposed the law, widespread debate and agitation took place across Mexican society. In the Mexican digital public sphere, the Twitter hashtag #LeyGeolocalización was widely used to discuss the law. Online communities like Human Rights Geek offered detailed analyses of the law and why they claimed it breached the legal framework of the Mexican Constitution. The law was also discussed under the hashtag #leystalker ("stalker law"), which emphasized the concerns over state surveillance.[28] Internationally, the Electronic Frontier Foundation called the law "alarming" and made the following statement:

> There is significant potential for abuse of these new powers. The bill ignores the fact that most cellular phones today constantly transmit detailed location data about every individual to their carriers; as all this location data is housed in one place—with the telecommunications service provider—police will have access to more precise, more comprehensive and more pervasive data than would ever have been possible with the use of tracking devices. The Mexican government should be more sensitive to the fact that mobile companies are now recording detailed footprints of our daily lives.[29]

The debate over the Geolocalization Law fed into other recent efforts to question the state's methods of surveillance using Internet, mobile, and locative technologies, along with the legal and regulatory frameworks concerning these. In July 2013, for instance, NGOs such as Propuesta Cívica and Contingente MX, asked the PGR for details on how the software Hunter Punta Tracking (geolocation) and FinFisher (which spies on Blackberry messages) are being used for surveillance purposes in federal investigations. Pilar Tavera from Propuesta Cívica summed up the mood saying: "No one has the right to search for data on your cellphone unless it is mandated by a judge."[30] In June 2013, Tavira also presented a claim to the IFAI (Instituto Federal de Acceso a la Información), asking them to investigate the carriers Iusacell and UniNet, which allegedly hosted FinFisher on overseas servers.[31] It is important to note that such opponents to this law argue not against the right of the state to curb criminality but against possible deviations of this right. This focus points to a lack of trust in institutions and public servants.

In response, Congressman José Adán Rubí Salazar, president of the Communications Commission of the House of Congress, defended the law, stating that "there are two extremes concerning risks; one damages people's privacy, the other one relates to impunity when dealing with criminals."[32]

For Rubí Salazar, the power given to the state to locate individuals through their mobile devices is fully justified.

On January 17, 2014, the Supreme Court ratified the Geolocalization Law as constitutionally sound, overturning the claims presented by the National Human Rights Commission. The final vote tally was 8 votes in favour of its ratification and 3 votes against. Minister Margarita Luna Ramos, who stated that the geolocalization of mobile devices did not violate a person's right to privacy and intimacy, championed the law.[33] Jesús Rodríguez Almeida, head of Mexico City's Public Security Secretariat (SSP-DF), supported the decision, stating that GPS systems have helped the authorities capture kidnappers and robbers by revealing their locations.[34]

On the other hand, ministers José Ramón Cossío Díaz, Olga Sánchez Cordero, and Sergio Valls Hernández opposed the law.[35] Valls Hernández stated that there was no guarantee that the attributions enabled by the law would not lead to abuses and that the court was acting "in good faith."[36] This view was echoed by a panel of academics from the Universidad Iberoamericana (UIA), one of Mexico's elite private universities. Human rights scholar Sandra Salcedo claimed that the ratification of the law was an "irresponsible" act and that the Supreme Court did not establish "clear mechanisms in which this sort of veiled surveillance can be put in place."[37]

The Supreme Court's ratification of the Gelocalization Law has not seen the controversy dissipate. If anything, the affirmation of the law has heightened the sense among many across Mexican society that it is a watershed moment.

CONCLUSION

By 2014, the technologies of locative media in smartphones and tablets were widely, if often very unequally, diffused across Mexico. If the craze for smartphones can be dated from the introduction of Apple's iPhone in 2007, then the technology has certainly been introduced in a very fraught context for Mexican politics, society, and everyday life. The last two presidencies (Felipe Calderón, 2006–2012, and Enrique Peña Nieto, 2012–2018) have seen an increase in the activity of federal and local security forces due to frontal, full-fledged state war against the drug cartels and other criminal networks.

Unsurprisingly, then, very early on in their diffusion in Mexico among elites and middle classes, such locative media have immediately resulted in an amalgam of anxieties, hopes, and fears that have formed a key element in accompanying social imaginaries. This culturally specific social imaginary of locative media can be seen in the development and debates accompanying the Geolocalization Law, our subject here. The ratification of the Geolocalization Law occurred as this book went to press, and little information, legal cases, or research is available concerning its actual implementation.

So, while the controversy continues unabated, it will be important—even if doubtlessly difficult for all sorts of reasons—to evaluate detailed information on how the law is put into effect across urban, regional, and rural areas, as well as the overlapping, conflictual layers and jurisdictions of the Mexican army, police, and judiciary.

For the wider international discussion, theorization, and research agenda, we would suggest that the case of the Mexican Geolocalization Law underscores the cultural and normative complexities generated by the use of locative media in particular national contexts. It also points to the new legal and political dynamics that sprout from the generation of and access to new types of data. Geolocative data has generated challenges to the cultural and legal notions of privacy because it is part of a complex technosocial network that involves individuals (citizens, voters, consumers), private companies (telecom carriers and device manufacturers), local government bodies, and international telecommunication agreements.

How to gain access to these different dimensions of locative media and personal information, in order to follow these actors across genuinely violent and dangerous landscapes—like those encountered in contemporary Mexico but also in many other countries—is quite a research challenge. Little wonder that researchers, leaders, officials, and civil society activists alike rely on the leaks, made famous by Wikileaks, Edward Snowden, and others. However, these adventitious sources of information come from breaches in the circuits of intelligence in the global north countries, while the economies and politics of personal information and mobile technology in the majority world of the global south develop apace, but not unchecked.

ACKNOWLEDGMENTS

Our thanks to Rodrigo Diez for his helpful comments and assistance. The chapter is an output of the Australian Research Council Discovery grant, *Moving Media: Mobile Internet and New Policy Modes* (DP120101971).

NOTES

1. Andrew Boulton and Matthew Zook, “Landscape, Locative Media, and the Duplicity of Code,” in *The Wiley-Blackwell Companion to Cultural Geography*, eds. Nuala C. Johnson, Richard H. Schein, and Jamie Winders (Malden, MA: Wiley, 2013), 442.
2. Notable studies include Barbara A. Crow, Kim Sawchuk, and Michael Longford, eds., *The Wireless Spectrum: The Politics, Practices, and Poetics of Mobile Media* (Toronto: University of Toronto Press, 2010); Eric Gordon and Adriana de Souza e Silva, *Net Locality: Why Location Matters in a Networked World* (Malden, MA: Wiley, 2011); Adriana de Souza e Silva and Jordan Frith, *Mobile Interfaces in Public Spaces: Locational Privacy, Control, and Urban Sociability* (London and New York: Routledge, 2012); Jason Farman, *Mobile*

Interface Theory: Embodied Space and Locative Media (London and New York: Routledge, 2012); Larissa Hjorth and Michael Arnold, *Online@AsiaPacific: Mobile, Social and Locative Media in the Asia-Pacific* (London and New York: Routledge, 2013).

3. Useful studies of Ushahidi as a participatory media platform include Ory Okolloh, "Ushahidi, or 'Testimony:' Web 2.0 Tools for Crowdsourcing Crisis Information," *Participatory Learning and Action* 59, no. 1 (2009): 65–70; D. Ndirangu Wachanga. "Participatory Culture in an Emerging Information Ecosystem: Lessons from Ushahidi," *Communicatio* 38, no. 2 (2012): 195–212; Janet Marsden, "Stigmergic Self-Organization and the Improvisation of Ushahidi," *Cognitive Systems Research* 21 (2013): 52–64.
4. Notably, the pioneering work of Adriana de Souza e Silva—for instance, her coauthored paper: Fernanda Duarte and Adriana de Souza e Silva, "Arte.mov, Mobilefest and the Emergence of a Mobile Culture in Brazil," in *The Routledge Companion to Mobile Media*, eds. Gerard Goggin and Larissa Hjorth (New York: Routledge, 2014), 206–215. See also André Lemos, "Locative Media in Brazil," *Wi: Journal of Mobile Media*, 3 (May 2009), http://wi.hexagram.ca/?p=60; various chapters in Rodrigo José Firmino, Fábio Duarte, and Clovis Ultramari, eds., *ICTs for Mobile and Ubiquitous Urban Infrastructures: Surveillance, Locative Media, and Global Networks* (New York: IGI Global, 2011).
5. On internationalizing research approaches, see Ackbar Abbas and John Erni, eds., *Internationalizing Cultural Studies: An Anthology* (Malden, MA: Blackwell, 2005); Gerard Goggin and Mark McLelland, eds., *Internationalizing Internet Studies: Beyond Anglophone Paradigms* (New York: Routledge, 2009); Daya Thussu, ed., *Internationalizing Media Studies* (New York: Routledge, 2009); Gerard Goggin and Mark McLelland, eds., *Routledge Companion to Global Internet Histories* (New York: Routledge, 2015).
6. Boulton and Zook note: "We should be careful not to make unsustainable, exaggerated, or ethnocentric claims about the ways in which locative media always, everywhere, and for everyone work" ("Landscape, Locative Media, and the Duplicity of Code," 442).
7. See, for instance, the discussion of Mexico in John Sinclair's classic study, *Latin American Television: A Global View* (Oxford and New York: Oxford University Press, 1999). More recently, see the Mexican case study in Anna Cristina Pertierra and Graeme Turner, *Locating Television: Zones of Consumption* (London and New York: Routledge, 2012).
8. ITU, "Mobile-Cellular Subscriptions," *Time Series*, accessed March 3, 2014, www.itu.int/en/ITU-D/Statistics/Pages/default.aspx.
9. *Mexico Telecommunications Report Q4 2013* (London: Business Monitor International, 2013).
10. Julio Vélez, "Top 10 de las apps mexicanas más exitosas," *Alto Nivel* (August 1, 2013), accessed March 3, 2014, www.altonivel.com.mx/37317-top-10-de-las-apps-mexicanas-mas-exitosas.html.
11. Pepe Flores, "¿Por qué tanta paranoia contra Foursquare en México?," *ALT1040* (April 1, 2011), accessed March 3, 2014, http://alt1040.com/2011/04/%C2%BFpor-que-tanta-paranoia-contra-foursquare-en-mexico.
12. Oliverio Pérez Villegas, "Taxibeat, una app de Grecia para la Ciudad de México," *Alto Nivel* (August, 22, 2013), accessed March 3, 2014, www.altonivel.com.mx/37582-taxibeat-la-revolucion-en-servicio-de-transporte.html.
13. WorldCrunch, "Mexico City Launches Smartphone App to Geolocate Nearest Cop," *WorldCrunch* (March 14, 2013), accessed March 3, 2014, www.worldcrunch.com/culture-society/mexico-city-launches-smartphone-app-to-geolocate-nearest-cop/mi-policia-mexico-safety-crime-police/c3s11186/.

14. Marguerite Cawley, "Mexico City Launches Crime-Fighting Smartphone App," *InSight Crime* (March 13, 2013), accessed March 3, 2014, www.insightcrime.org/news-briefs/mexico-city-smartphone-app-fighting-crime.
15. Jens Glüsing, Laura Poitras, Marcel Rosenbach, and Holger Stark, "Fresh Leak on US Spying: NSA Accessed Mexican President's Email," *Der Spiegel* (October 20, 2013), accessed March 3, 2014, www.spiegel.de/international/world/nsa-hacked-email-account-of-mexican-president-a-928817.html.
16. Kathleen Hennessey and Vincent Bevins, "Brazil's President, Angry About Spying, Cancels State Visit to U.S.," *Los Angeles Times* (September 17, 2013), accessed March 3, 2014, http://articles.latimes.com/2013/sep/17/world/la-fg-snowden-fallout-20130918.
17. Jorge A. Vargas, "Privacy Rights Under Mexican Law: Emergence and Legal Configuration of a Panoply of New Rights," *Houston Journal International Law* 27 (2004): 76–77.
18. Vargas, "Privacy Rights Under Mexican Law," 82.
19. Pepe Flores, "La Suprema Corte de México debate la polémica Ley de Geolocalización," *Fayer Wayer* (January 9, 2014), accessed March 3, 2014, www.fayerwayer.com/2014/01/la-suprema-corte-de-mexico-debate-la-polemica-ley-de-geolocalizacion/.
20. Raquel Seco, "Las prisiones mexicanas se especializan en extorsión telefónica," *El País* (July 7, 2013).
21. Alejandro Martí, "¿Qué es la Ley de Geolocalización?" *Animal Político* (March 6, 2012), accessed March 3, 2014, www.animalpolitico.com/blogueros-mexico-sos/2012/03/06/que-es-la-ley-de-geolocalizacion/#axzz2rvw7nVuA.
22. Senador Tomás Torres Mercado, speech, *LXII Legislatura, Senate of the Republic of Mexico*, versión stenográfica (Wednesday, March 15, 2011), accessed March 3, 2014, www.senado.gob.mx/index.php?ver=sp&mn=4&sm=2&fecha=2011–3-15.
23. Torres Mercado, speech.
24. Torres Mercado, speech.
25. Leonardo Peralta, "La 'ley de Geolicalización' permitirá 'rastear' a usuarios con celular," *CNN México* (March 27, 2012), accessed March 3, 2014, http://mexico.cnn.com/tecnologia/2012/03/27/la-ley-de-geolocalizacion-permitira-rastrear-a-usuarios-de-celular.
26. *Gaceta Parlamentaria*, year XV, number 3462-III (March 1, 2012), accessed March 3, 2014, www.diputados.gob.mx/cedia/sia/dir/DIR-ISS-04–12_anexo_dic1.pdf. Translation by the authors.
27. Mariana Alviso, "¿Ley de geolocalización o ley stalker?" *Interactive Magazine* (April 25, 2012), accessed March 3, 2014, http://revistainteractive.com/tecno-geolocalizacion-stalker/.
28. Mariana Alviso, "¿Ley de geolocalización o ley stalker?" *Interactive Magazine* (April 25, 2012), accessed March 3, 2014, http://revistainteractive.com/tecno-geolocalizacion-stalker/.
29. Katitza Rodriguez, "Mexico Adopts Alarming Surveillance Legislation," Electronic Frontier Foundation (March 2, 2012), accessed March 3, 2014, https://www.eff.org/deeplinks/2012/03/mexico-adopts-surveillance-legislation.
30. Pilar Tavera, quoted in Diana Baptista, "Piden ONG aclarar uso de software," *Mural* (July 7, 2013), accessed March 3, 2014, http://mural-guadalajara.vlex.com.mx/vid/piden-aclarar-uso-spyware-446690942.
31. Baptista, "Piden ONG aclarar uso de software."
32. Notimex, "Ley de geolocalización no vulnera derechos humanos: Rubí Salazar," *Notimex: Agencia Mexicana de Noticias* (May 25, 2012), accessed

March 3, 2014, http://mx.noticias.yahoo.com/ley-geolocalizaci%C3%B3n-vulnera-derechos-humanos-rub%C3%AD-salazar-191600306.html.

33. Isabel González, "SCJN discute constitucionalidad de geolocalización," *Excélsior* (January 9, 2014), accessed March 3, 2014, www.excelsior.com.mx/nacional/2014/01/09/937443.
34. Alejandro Cruz, "Respalda SSPDF aval de Corte sobre geolocalización de celulares," *La Jornada* (January 16, 2014), accessed March 3, 2014, www.jornada.unam.mx/ultimas/2014/01/16/respalda-sspdf-aval-de-corte-sobre-geolocalizacion-de-celulares-4911.html.
35. Jesús Aranda, "Continúa intenso debate en la Corte sobregeolocalización de celulares," *La Jornada* (January 14, 2014), accessed March 3, 2014, www.jornada.unam.mx/ultimas/2014/01/14/continua-debate-en-la-corte-sobre-si-debe-limitar-geolocalizacion-de-celulares-1360.html.
36. González, "SCJN declara constitucional la geolocalización."
37. Notimex, "Expertos ven necesario fijar leyes claras para ley de geolocalización," *Noticias MVS* (January 30, 2014), accessed March 3, 2014, http://noticiasmvs.com/#!/noticias/expertos-ven-necesario-fijar-reglas-claras-para-ley-de-geolocalizacion-613.html.

12 Seeking Transparency in Locative Media

Tama Leaver and Clare Lloyd

INTRODUCTION

A person's location is, by its very nature, ephemeral, continually changing and shifting. Locative media, by contrast, entails processes in which a user's geographic location is encoded, usually along with an exact time stamp, translating this data into information that not only persists but can be aggregated, searched, indexed, mapped, analyzed, and recalled in a variety of ways for a range of purposes.[1] However, even though the utility of locative media for the purposes of tracking, advertising, and profiling is obvious to many large corporations, these uses are often far from clear for many users of mobile media devices such as smartphones, tablets, and satellite navigation tools. Moreover, when a new mobile media device is purchased, users are often overwhelmed with the sheer number of options, tools, and apps at their disposal. Exploring the settings or privacy preferences of a new device in a sufficiently granular manner to even notice the various location-related options can understandably escape many users.

As Lane DeNicola has argued, corporate slogans such as Apple's "It Just Works" aptly summarize a culture of occlusion whereby consumers are positioned as purchasing the right to the surface of a device or program but are actively discouraged from exploring their inner working, be that hardware or software.[2] Similarly, even users who initially deactivate geolocation tracking often unintentionally reactivate it and leave it on, in order to use the full functionality of many apps. For social apps that utilize and generate locative media, a *social media contradiction* may arise whereby users focus on the *social* elements—often the acts of communication and sharing, which are thought of as ephemeral and in the moment, comparable to a telephone conversation—while the companies and corporations creating these apps are more focused on the *media* elements, which are measurable, aggregatable, can be algorithmically analyzed in a variety of potentially valuable ways and which can last indefinitely.[3] While concepts of privacy and the presumed boundary between public and private are challenged and extended by social and locative media, users need to be aware of the amount of information they are sharing to actively participate in this ongoing (re)negotiation.[4]

In this chapter, we explore pedagogical practices aimed at broadening the awareness of information created while using locative media, with the aim of overcoming the social media contradiction.

This chapter will initially outline how the meaning of transparency has shifted in relation to technology, once signifying that devices were open to tinkering at the level of hardware and software but more recently being deployed to mark devices and their workings as seemingly invisible, to be looked *through*, not at. Secondly, the ways in which locative media blur the boundaries between public and private will be explored, in tandem with the related division between surveillance and privacy. The following section will look at purposeful techniques and pedagogical strategies that can be deployed to increase the awareness of locative media and related metadata, including an examination of four tools—Please Rob Me, I Can Stalk U, iPhone Tracker, and Creepy[5]—all of which enact this pedagogical practice. Finally, the chapter will conclude by situating the challenges concerning users' knowledge of locative media as part of a broader social media contradiction, which needs to be overcome to ensure that all users can make informed choices about the exchanges they are entering into when creating and sharing location-based information using tools designed by private, usually for-profit corporations.

FROM TRANSPARENCY TO OCCLUSION

Author and futurist Arthur C. Clarke famously argued that "any sufficiently advanced technology is indistinguishable from magic,"[6] but for many users of locative media, today it is the devices in their hands that appear to be powered by magic, although this was not always the case. In an insightful exploration of design principles over the lifetime of the Apple corporation, Lev Manovich argues that the initial openness and accessibility of the early Macintosh computers have been superseded by what he calls an "aesthetic of disappearance" starting from the 1998 introduction of iMac G3 and its clear, colored cases and continuing today at both hardware and software levels: "Both the translucency of iMac's plastic case, the Dock magnification and Genie effects in the Aqua interface, and the iPhone's iOS all similarly stage technology as magical and almost supernatural."[7] Moreover, Manovich argues that, driven by Apple chief designer Jonathan Ive's aesthetics, "the technological object wants to disappear, fade into the background, and become ambient rather than actively attracting attention to itself and its technological magic like the first iMac."[8] The iPhone and iPad are thus the epitome of a tandem software and hardware design process that attempts to erase the device from view altogether. Rather than point to themselves, these devices at first glance appear to be frames, sporting increasingly dense and sizable screens, more like a painting than a desktop computer. These devices are literally transparent in DeNicola's sense that, in contemporary use, "a 'transparent' device is one whose use is highly intuitive, employing an interface that rapidly disappears from our perception. In the earlier meaning of the term, a 'transparent' device was one whose inner workings were visible and accessible, a device amenable to functional understanding and rapid appropriation."[9]

As Larissa Hjorth and Ingrid Richardson note, "the rise of mobile media is characterised by the rise of the active and creative user,"[10] and yet this active, creative user can be seen as restrained in quite explicit ways. Jonathan Zittrain, for example, situates the shift from Apple's Macintosh to the iPhone and similar devices in the early twenty-first century as a lamentable movement from generative to tethered technologies. Generative technologies are, for example, personal computers, which allow users to produce their own change, to write and run their own code, share that code without restrictions, and, in general, generate new programming without any outside interference. Tethered technologies, by contrast, like the iPhone, have a limited capacity for purely user-driven change and creativity because users cannot access the code level of the devices themselves. To create, run, and distribute code for tethered technologies, users usually have to get permission from at least one external gatekeeper.[11] In the case of the iPhone, the gatekeeper is Apple's App store. However, it is worth noting that Zittrain's dichotomy is about the potential of a device; in real terms, the vast majority of people using personal computers never actually explore at the level of programming or code, so for the average users the differences between tethered and generative technologies are more political and philosophical than experiential.

At the software rather than at the hardware level, Gerard Goggin argues that apps *en masse* are both ubiquitous and bring new challenges in conceptualizing their importance in the mobile media ecosystem.[12] While definitely epicenters for innovation, each significant mobile operating system has its own app store, with varying degrees of oversight, from Apple's very heavy-handed rules regarding apps through to Google's Play store, which at least attempts to be more open, relying on user reviews rather than on central policy, although it too reserves the right to remove apps for various reasons.[13]

The various competing mobile operating systems and the ecologies of apps within them are also wrapped in layers of obfuscation. End user license agreements (EULAs), for example, are usually written in language so lengthy and impenetrable that, despite governing the use of a piece of software, it almost never gets read.[14] Along very similar lines, Adriana de Souza e Silva and Jordon Frith argue that, in terms of locative services, just like online services more generally, "users agree to privacy policies that are nearly incomprehensible."[15]

Both EULAs and privacy policies further distance users from a sense that they can control or even understand how their mobile devices and apps operate. DeNicola characterizes the design of the iPhone, other touch screen mobile devices, and the software and apps they run, as producing an "occluded culture"[16] wherein deeper meaning and workings are entirely hidden. Within this context, any attempt by users to understand locative media must already combat multiple levels of occlusion in which mobile devices themselves, as well as the specific apps and legal terms they are wrapped in, all deflect attention away from themselves, encouraging users to accept that, as the Apple slogan celebrates, "It just works."

BEYOND PRIVACY AND OPTING OUT

Even though one response to concerns about social media and privacy is simply to advocate that users do not share information publicly, the line between public and private blurs when mobile and locative media are taken into account. Earlier concepts of privacy were often based on the more rigid demarcation of physical spaces into private (exemplified by the domestic home) and public, yet mobile and locative media perforate this division, meaning that the most intimate private conversations might be happening in the most public of locations (such as arguing with a loved one on a mobile phone while on a crowded train) or vice versa.[17] As Hjorth and Richardson argue, "the domestic, private, and personal become quite literally *mobilised* and *micro-mediatised* via the mobile phone—an intimate 'home-in-the-hand'—effecting at the same time a transformation of experiences of presence, telepresence, and co-presence in public spaces."[18] When the private and personal are mediated through a mobile device, this has the potential to recode notions of public and private in ways that challenge a clear or consistent division between the two. Of course, as de Souza e Silva and Frith quite rightly note, far from concrete or static notions, "neither public nor private are objective entities: They are socially negotiated and constantly shifting."[19] In light of locative and mobile media, the terms "public" and "private" thus become "less and less useful to accurately describe our experiences,"[20] and yet, due to their pervasiveness, these remain the terms against which so much activity is measured. Moreover, as the notion of private becomes more fluid, the question of privacy and an individual's right to it also become problematic.

As legal scholar Julie Cohen notes, although there is a scholarly consensus that notions of privacy are still important and are changing, there is absolutely no consensus on how privacy laws, rules, and norms should actually operate today in light of mobile devices and other recent developments.[21] Moreover, as Cohen argues further, participating less in social, mobile, and locative media might increase privacy but at a potentially very high cost in terms of participation: "[c]hoices about privacy are choices about the scope for self-articulation."[22] Illustrating that simply opting out is far from a straightforward proposition, it is noteworthy that, as early as 2011, one in three Google searches from mobile services directly related to the current location of the user.[23] Increasing privacy by avoiding locative media does not simply mean not geotagging Instagram photos or avoiding the locative play of Foursquare; it entails limiting search options, avoiding online maps on mobile devices, not getting weather updates, and not even checking the traffic conditions before driving home.

In 2006, Susan Barnes argued that a "privacy paradox" existed in the United States where social media tools were leading to unintended public sharing because many users (teenagers, in Barnes' case study) were unaware of the public nature of much of their online communication.[24] Even though far more users today are aware that online information is likely public unless otherwise specified—in part because the architecture of the major social media platforms has shifted from private by default to public by default—the question of locative media remains complicated. In part, this is because of not just the surface level of locative media devices but also locative metadata, such as geotags (geographic tags) tied to photographs or social media updates, especially since so many people have no idea these exist. That said, it is certainly not the case that all users are unaware of locative media and metadata.

In August 2013, the Pew Research Centre's Internet and American Life Project released a report based on a nationally representative sample of 802 U.S. teens aged 12 to 17 years that found that, far from ignorant of locative media, 46 percent of those teens who had ever downloaded an app said that they had turned off location-tracking features either on the phone overall or within a specific app. There was a significant gender divide, with only 37 percent of boys disabling location tracking, whereas 59 percent of girls had done so. Privacy appears to be the key concern, yet the report noted that the survey questions did not discriminate between those who were concerned about corporations or parents having access to their location-based information.[25] De Souza e Silva and Frith identify three types of locative surveillance: governmental surveillance (government gaining access to locative information); corporate surveillance (advertisers and marketers exploiting locative data to better target advertising); and collateral surveillance (users' unintended sharing of locative data with other users).[26] Governmental and corporate surveillance are clearly both top-down processes, but collateral surveillance is a comparably newer form and includes the realm of parents tracking their children's movements or more playful versions, such as the use of Foursquare to find new locations with people you know already in them.

Surveillance itself is neither monolithic nor straightforward. Some critics see collateral surveillance—the ability of all locative media users to survey one another's activities—as a destabilizing kind of surveillance and as a necessarily oppressive practice. Anders Albrechtslund, for example, articulates the idea of "social surveillance," describing the way surveillance might be reconfigured as a form of knowing play,[27] whereas E. J. Westlake goes a step further, arguing that just because a user's online performance and activity are being recorded, there is nothing preventing these performances from being entirely false or misleading, subverting the effectiveness of even top-down surveillance.[28] Following Gordon and de Souza e Silva, the question of surveillance and the disclosure of location data "only really becomes a problem when users are unaware of who owns the information about their location, and with whom their location is being shared."[29] However, they continue to argue that for the majority of users, privacy settings on location-based services are hard to understand, can change with minimal or no notice, and often never come to a user's attention at all.[30] With this context in mind, the next section argues for the importance of tools designed to increase users' awareness of both how locative media and metadata work and how to contemplate fine-grained control over these services, not just imagining a gigantic on/off switch for privacy or surveillance.

RESTORYING USER AWARENESS

In *The Uses of Digital Literacy*, John Hartley argues that "not enough 'critical' attention has been paid to what ordinary people need to learn in order to attain a level of digital literacy appropriate for *producing* as well as *consuming* digital content, thence to participate in the mediated public sphere."[31] Hartley argues that young people are often failed by schools and teachers who are unable or unwilling to teach digital literacies, meaning that most young people acquire their creative and critical digital skills in an ad hoc manner at best, often leading to very purposive but not always comprehensive knowledge of the skills and tools needed for creative, critical, and civic expression.[32] At a fairly banal level, this may mean that users of a service like Facebook or Foursquare are well equipped to communicate using a variety of tools on each platform but are either unaware or uninterested in how any act of communication may also be creating a sea of recordable data and metadata, including quite often locative metadata. However, opting out of these services would mean opting out of the increasingly normalized and banal infrastructures of social communication, as well as personal and political activity.[33] Instead, this section analyzes four tools that have emerged that try to use innovative pedagogical techniques to increase the general public's awareness of how locative media services operate. These tools resonate with the arguments put forward in *Spreadable Media: Creating Value and Meaning in a Networked Culture* by Henry Jenkins, Sam Ford, and Joshua Green, in which they examine some of the methods by which different types of media and messages gain widespread traction. They identify a number of elements that can increase the spread of a message, including media forms that are "unfinished" in the sense that users need to do something to make the most of that artifact; media that include an element of mystery, needing to be engaged with in order to reveal something already alluded to but not yet clear; and, especially, media that engage with a timely controversy because these forms speak about our immediate and pressing everyday concerns.[34] While spreadable media are often commercial, the same principles can effectively apply to educational and political messages: "spreadable civic media content may be initially jarring in the ways that it abandons the sobriety with which we normally receive political messages, but producers count on the controversy around such unexpected tactics to inspire further spread and discussion."[35] Moreover, as Rowan Wilken notes, media controversies over locative media have been one of the key drivers increasing public awareness of the way particular tools, services, and platforms generate and store geographically tagged data and metadata.[36]

Please Rob Me and I Can Stalk U are both web-based educational tools that were launched in 2010 with the aim of illustrating how locative media metadata might be (mis)used. In essence, Please Rob Me simply displayed a customized Twitter search; each time a message from the locative service Foursquare was pushed to Twitter, it was displayed on the Please Rob Me web page. Tweets were reformatted and displayed "@username left home and checked in less than a minute ago: I'm at [placename] [http://urlshowinglocation.com]." The point was simply to highlight that every time people publicly shared their locations, they implicitly also stated that they were *not* at home and that their home might thus be a good target for burglary. This was especially troubling for people who had publicly shared the address of their home, either by checking in to it themselves or by having friends who had done so, thus creating a public record of their home address. As the Please Rob Me website explains, the service was meant as a satire with an educational intent:

> The danger is publicly telling people where you are. This is because it leaves one place you're definitely not . . . home. So here we are; on one end we're leaving lights on when we're going on a holiday, and on the other we're telling everybody on the internet we're not home. It gets even worse if you have "friends" who want to colonize your house. That means they have to enter your address, to tell everyone where they are. Your address . . . on the internet . . . Now you know what to do when people reach for their phone as soon as they enter your home.[37]

Please Rob Me actively courted controversy and gained significant mainstream press attention in the process.[38]

Several months later, I Can Stalk U was released, which was similar to Please Rob Me in that it repurposed metadata and made that the main item displayed. I Can Stalk U displayed any tweet that included a photograph with a geotag (metadata in image files that shows the precise latitude and longitude coordinates of where the photograph was taken) in the format "@username: I am currently at [exact address]"; for example "tamaleaver is currently near 35 Stirling Highway, Crawley, Western Australia." Each post included the ability to see the photograph, the original tweet, the exact geographic coordinates on a map, or reply to the user who originally made the tweet. The I Can Stalk U website also included a clear explanation of the educational rationale for the tool; the designers simply wanted to highlight the way geotags on photographs operate. More importantly, on the How? page the designers wrote a very clear document explaining what metadata is, what geotags are, and how they are generated; most importantly, the website contained a thorough list of the ways in which users could disable automatic geotagging of their photographs on most popular mobile devices. These two awareness-raising tools actively courted controversy both with their names and in how they displayed repurposed locative metadata, with Please Rob Me demonstrating the privacy concerns with Foursquare check-ins and with I Can Stalk U highlighting the existence and impact of geotags on publicly shared photographs. Notably, the negative publicity generated was sufficient that Foursquare posted an official response to Please Rob Me, noting the company's commitment to privacy while reminding their users that "location is sensitive data and people should be careful about with whom and when they share it."[39]

When Alasdair Allan and Pete Warden discovered that iOS 4 (the then current operating system of the iPhone and 3G-enabled iPads) was keeping a log of location data, including a potentially yearlong record of each and every wireless tower accessed, it provoked a storm of controversy and staunch criticism of Apple.[40] As Gerard Goggin notes, Apple immediately positioned the length of the log file as a bug, while arguing that a more short-term log was simply intended to help speed up and improve the accuracy of location-based services.[41] More importantly for the argument in this chapter, Allan and Warden also publicly released a tool called iPhone Tracker, which allowed any iPhone user to easily access and visualize the log file created by their own iPhone.[42] The result was a map showing the locations visited over a substantial period of time, evoking an eerie sense that the iPhone was spying on its owner. Kate Crawford notes that this level of data was always available to mobile phone carriers, but the accessibility of this data to everyday users provoked a significant privacy scandal for Apple. The iPhone Tracker raised many people's awareness of the sort of longevity that location-based data can have.[43] Importantly, in providing a tool that any iPhone user could use, Allan and Warden engaged users' curiosity about their own trails and in doing so provoked further contemplation of the sort of information anyone's mobile devices might be collecting. As Crawford notes, even once the bug was fixed, "the concerns about location data on the iPhone iOS 4 nonetheless reveal the way these devices and the bargains we make with them can have unintended and long-lasting consequences."[44] Notable, too, was that, in Apple's official response to the log file and iPhone Tracker tool, the company explicitly commented that they and most other technology companies had failed to educate their users sufficiently about the operation of locative media on the devices they manufacture.[45]

The final tool of note is Ioannis Kakavas's Creepy.[46] Released in 2011, using a desktop interface, Creepy can aggregate locative metadata from Twitter, Foursquare, and the available geotags from a range of image-sharing tools, including Flickr, TwitPic, and YFrog. This record is then overlaid on a map that can be navigated temporally, with the scroll of a timeline showing the places and times of a person's visits if he or she is wielding a locative media device. Having a map that can show a user's movements over time can be quite disconcerting if they were unaware that this metadata is being recorded. Similar to the other tools mentioned, Creepy's creator explicitly positions it as an educational tool:

> Well, I don't think that the fact that your geolocation information can be gathered and aggregated is disturbing. The fact that you were publishing it in the first place, is, on the other hand. Just to be clear, the intention behind creating Creepy was not to help stalkers or promote/endorse stalking. It was to show exactly how easy it is to aggregate geolocation information and make you think twice next time you opt-in for geolocation features in twitter, or hitting "allow" in the "this application wants to use your current location" dialog on your iPhone.[47]

Although Creepy is the most expansive of the tools examined, drawing on multiple services, it nevertheless shares the same aims of raising awareness and courting controversy to increase overall attention to how locative media works.

All such awareness-raising tools avoid any explicit wholesale denunciation of locative media, favoring instead approaches that highlight the nuances and potential consequences of specific types of use. Using the initial guise of moral finger wagging, Please Rob Me and I Can Stalk U both courted user and media attention but then used that interest to educate users about the potential challenges and ramifications of Foursquare check-ins and geotagged photographs, respectively. iPhone Tracker allowed curious users to extract and explore their own version histories as tracked on iOS devices, whereas Creepy goes a step further and allows the mapping of a range of locative media tools. Unsurprisingly, all of these tools gained mainstream media attention, but, significantly, the documentation provided by Please Rob Me, iPhone Tracker, and Creepy all explain that these are awareness-raising tools, designed to help everyday users appreciate how much information they may be unintentionally sharing, as well as how that information could be harnessed in other contexts. I Can Stalk U goes a step further with explicit instructions on how to disable the embedding of geotags on a range of smartphones. These tools offer different narratives, restorying the same data and metadata to tell quite different and provocative tales. In terms of locative media, these tools are assisting everyday users in developing the literacies needed to utilize mobile devices and locative media tools in an informed and purposeful way, without resorting to an all-or-nothing approach to utilize location-based tools.

CONCLUSION

As the era of wearable computing looms large, complete with its first major icon, Google's wearable Glasses (see Chapter 10), the potential for the interface to disappear altogether is all too real. While mobile and locative media are at the core of wearable computing, the challenge remains to make the operation of these media understandable and visible to everyday users today, before they are wrapped in yet another layer of invisibility and obfuscation. A social media contradiction—where users generally focus on immediate *social* uses such as communication, while the corporations behind these tools focus more on the ongoing value of the *media* and data generated by user activity—occurs only when the longevity and utility of data and information generated by social, mobile, or locative media are occluded, hidden from the view of everyday users. Although Foursquare appeared for some years to be lacking a clear business model, for example, the immense value of the locative data that the users of the app have generated was made explicit by a bidding war in August 2013 that saw Yahoo!, Apple, and Microsoft all vying for access to Foursquare's entire locative media database.[48] The awareness-raising tools analyzed in this chapter are an important part of a larger pedagogic strategy to ensure that users have the opportunity to be equally aware of the value of the locative media they generate. The media coverage and online discussion generated by Please Rob Me, I Can Stalk U, iPhone Tracker, and Creepy all brought the question of locative media into the spotlight, increasing user awareness of locative media and metadata. The websites housing these tools all highlighted their educational aims, while several explicitly detailed techniques for minimizing or more carefully controlling the generation of locative metadata on specific devices. The ensuing public attention was sufficient to provoke official responses, including those from Foursquare and Apple, which both reminded users that they needed to be aware of how locative media operate and explicitly acknowledged that more education was needed for users (albeit without actually offering to provide material to further that education). The awareness-raising tools analyzed in this chapter form an important part of efforts to make locative and other media forms more visible and less mysterious or magical. Further work is sorely needed in this area. Governments, advocacy groups, and educators would be well served in noting the provocative strategies used by these awareness-raising tools in successfully capturing the attention of the media. Such strategies could inform future digital literacy efforts in overcoming the social media contradiction in relation to social and locative media. The rich potential of locative media can be better understood only if users, educators, and the companies themselves seek to make locative media transparent, not in the sense of looking through purposefully invisible devices but rather in making the operation, storage, aggregation, and analysis of the data generated by using locative media a process that is visible and evident to any interested user.

NOTES

1. Kate Crawford, "Four Ways of Listening with an iPhone: From Sound and Network Listening to Biometric Data and Geolocative Tracking," in *Studying Mobile Media: Cultural Technologies, Mobile Communication, and the iPhone*, eds. Larissa Hjorth, Jean Burgess, and Ingrid Richardson (London and New York: Routledge, 2012), 213–228; Alison Gazzard, "Location, Location, Location: Collecting Space and Place in Mobile Media," *Convergence: The International Journal of Research into New Media Technologies* 17, no. 4 (November 2011): 405–417, doi:10.1177/1354856511414344.
2. Lane DeNicola, "EULA, Codec, API: On the Opacity of Digital Culture," in *Moving Data: The iPhone and the Future of Media*, eds. Pelle Snickars and Patrick Vonderau (New York: Columbia University Press, 2012), 265–277; see also Jonathan Zittrain, *The Future of the Internet—And How to Stop It* (New Haven and London: Yale University Press, 2008).
3. Tama Leaver, "The Social Media Contradiction: Data Mining and Digital Death," *M/C Journal* 16, no. 2 (May 2013), http://journal.media-culture.org.au/index.php/mcjournal/article/viewArticle/625.
4. danah boyd, "Facebook's Privacy Trainwreck: Exposure, Invasion, and Social Convergence," *Convergence* 14, no. 1 (2008): 13–20; Adriana de Souza e Silva and Jordan Frith, *Mobile Interfaces in Public Spaces: Locational Privacy, Control, and Urban Sociability* (London and New York: Routledge, 2012); Jason Farman, *Mobile Interface Theory: Embodied Space and Locative Media* (New York and London: Routledge, 2012); Maren Hartmann, "Mobile Privacy: Contexts," in *Privacy Online: Perspectives on Privacy and Self-Disclosure in the Social Web*, eds. Sabine Trepte and Leonard Reinecke (Heidelberg: Springer, 2011), 191–203; Bridgette Wessels, "Identification and the Practices of Identity and Privacy in Everyday Digital Communication," *New Media & Society* 14, no. 8 (December 1, 2012): 1251–1268, doi:10.1177/1461444812450679.
5. Barry Borsboom, Boy van Amstel, and Frank Groeneveld, *Please Rob Me*, accessed August 22, 2013, http://pleaserobme.com/; Ben Jackson, Larry Pesce, and Mayhemic Labs, *I Can Stalk U*, accessed August 22, 2013, http://ican stalku.com/; Ioannis Kakavas, *Creepy*, accessed August 22, 2013, http://ilek trojohn.github.com/creepy/; Alasdair Allan and Pete Warden, *iPhone Tracker*, accessed August 22, 2013, http://petewarden.github.io/iPhoneTracker/.
6. Arthur C Clarke, "Hazards of Prophecy: The Failure of Imagination," in *Profiles of the Future: An Inquiry into the Limits of the Possible* (New York: Harper & Row, 1973).
7. Lev Manovich, "The Back of Our Devices Look Better Than the Front of Anyone Else's: On Apple and Interface Design," in *Moving Data: The iPhone and the Future of Media*, eds. Pelle Snickars and Patrick Vonderau (New York: Columbia University Press, 2012), 284.
8. Manovich, "The Back of Our Devices," 285.
9. DeNicola, "EULA, Codec, API," 267.
10. Larissa Hjorth and Ingrid Richardson, "The Waiting Game: Complicating Notions of (Tele)presence and Gendered Distraction in Casual Mobile Gaming," *Australian Journal of Communication* 36, no. 1 (2009): 25.
11. Zittrain, *The Future of the Internet—And How to Stop It.*
12. Gerard Goggin, "Ubiquitous Apps: Politics of Openness in Global Mobile Cultures," *Digital Creativity* 22, no. 3 (2011): 148–159, doi:10.1080/146262 68.2011.603733.
13. Goggin, "Ubiquitous Apps," 154.

14. DeNicola, "EULA, Codec, API," 269.
15. de Souza e Silva and Frith, *Mobile Interfaces in Public Spaces*, 128.
16. DeNicola, "EULA, Codec, API," 268.
17. Hartmann, "Mobile Privacy: Contexts," 193–194.
18. Hjorth and Richardson, "The Waiting Game," 26.
19. de Souza e Silva and Frith, *Mobile Interfaces in Public Spaces*, 15.
20. Farman, *Mobile Interface Theory*, 71.
21. Julie E. Cohen, *Configuring the Networked Self: Law, Code, and the Play of Everyday Practice* (New Haven: Yale University Press, 2012), 108.
22. Cohen, *Configuring the Networked Self*, 149.
23. Carlos Barreneche, "Governing the Geocoded World: Environmentality and the Politics of Location Platforms," *Convergence: The International Journal of Research into New Media Technologies* 18, no. 3 (August 1, 2012): 340, doi:10.1177/1354856512442764.
24. Susan B. Barnes, "A Privacy Paradox: Social Networking in the United States," *First Monday* 11, no. 9 (2006), accessed August 22, 2013, http://firstmonday.org/ojs/index.php/fm/article/view/1394.
25. Mary Madden, Amanda Lenhart, Sandra Cortesi, and Urs Gasser, "Teens and Mobile Apps Privacy," *Pew Research Center*, accessed August 22, 2013, http://pewinternet.org/Reports/2013/Teens-and-Mobile-Apps-Privacy.aspx.
26. de Souza e Silva and Frith, *Mobile Interfaces in Public Spaces*, 121–127.
27. Anders Albrechtslund, "Online Social Networking as Participatory Surveillance," *First Monday* 13, no. 3 (March 2008), http://firstmonday.org/htbin/cgiwrap/bin/ojs/index.php/fm/article/view/2142/1949.
28. E. J. Westlake, "Friend Me If You Facebook: Generation Y and Performative Surveillance," *TDR: The Drama Review* 52, no. 4 (2008): 21–40.
29. Eric Gordon and Adriana de Souza e Silva, *Net Locality: Why Location Matters in a Networked World* (Oxford: Wiley-Blackwell, 2011), 138.
30. Gordon and de Souza e Silva, *Net Locality*, 138–140.
31. John Hartley, *The Uses of Digital Literacy* (St. Lucia: University of Queensland Press, 2009), 12.
32. Hartley, *The Uses of Digital Literacy*, 30.
33. Hartley, *The Uses of Digital Literacy*, 139.
34. Henry Jenkins, Sam Ford, and Joshua Green, *Spreadable Media: Creating Value and Meaning in a Networked Culture* (New York and London: New York University Press, 2013), 209–216.
35. Jenkins, Ford, and Green, *Spreadable Media*, 220.
36. Rowan Wilken, "Locative Media: From Specialized Preoccupation to Mainstream Fascination," *Convergence: The International Journal of Research into New Media Technologies* 18, no. 3 (August 2012): 243, doi:10.1177/1354856512444375.
37. Barry Borsboom, Boy van Amstel, and Frank Groeneveld, *Please Rob Me*, accessed August 22, 2013, http://pleaserobme.com/.
38. de Souza e Silva and Frith, *Mobile Interfaces*, 127; Wilken, "Locative Media": 243.
39. Foursquare, "On Foursquare, Location & Privacy . . . ," *The Foursquare Blog* (2010), accessed August 22, 2013, http://foursquare.tumblr.com/post/397625136/on-foursquare-location-privacy.
40. Alasdair Allan and Pete Warden, "Got an iPhone or 3G iPad? Apple Is Recording Your Moves," *O'Reilly Radar* (2011), accessed August 22, 2013, http://radar.oreilly.com/2011/04/apple-location-tracking.html.
41. Gerard Goggin, "Encoding Place: The Politics of Mobile Location Technologies," in *Mobile Technology and Place*, eds. Rowan Wilken and Gerard Goggin (London and New York: Routledge, 2012), 205.

42. Allan and Warden, *iPhone Tracker*.
43. Crawford, "Four Ways": 223–225.
44. Crawford, "Four Ways": 226.
45. Apple, "Apple Q&A on Location Data" (April 27, 2011), 2013, www.apple.com/pr/library/2011/04/27Apple-Q-A-on-Location-Data.html.
46. Ioannis Kakavas, *Creepy*, accessed August 22, 2013, http://ilektrojohn.github.com/creepy/.
47. Ioannis Kakavas, "FAQ," *Creepy*, accessed August 22, 2013, http://ilektrojohn.github.io/creepy/faq.html.
48. Austin Carr, "Why Yahoo and Apple Want Foursquare's Data," *Fast Company*, accessed August 23, 2013, www.fastcompany.com/3016250/why-yahoo-and-apple-want-foursquares-data; Sarah Frier, Douglas MacMillan, and Dina Bass, "Microsoft to Invest in Foursquare: Report," *The Age* (August 30, 2013), www.theage.com.au/it-pro/business-it/microsoft-amex-fighting-for-foursquare-report-20130830-hv1lc.html.

Part 4

Economies, Networks, Logistics

13 Locating Foursquare
The Political Economics of Mobile Social Software

Rowan Wilken and Peter Bayliss

INTRODUCTION

> [W]e must locate the Internet within the evolving media economy. We must learn to see how it fits within, and how it modifies, an existing forcefield of institutional structures and functions.[1]

In this chapter, we wish to take this quote from Dan Schiller as a point of departure for thinking about the development of the location-based mobile social software application Foursquare. One of the key aims of this chapter is to examine the economic dimensions of Foursquare in order to discover how it fits within the "existing forcefield" of "new media" corporate and economic structures and functions, as well as what its place might reveal about the ongoing evolution of commercial location services. To date, Foursquare has pursued integration with and between dominant new media platforms, such as Facebook and Twitter, and partnerships with established brands in other sectors, employing the principle of user-generated content (UGC) not only to provide content but also to develop key features and aspects of their service. In seeking to become *the* location database layer of the Internet and thus control the associated data sets, Foursquare has followed a precarious path reliant on partnerships with potential and actual competitors and on the contributions of both content and features from early adopters of the service.

Foursquare is a location-based mobile media service driven by check-ins to locations that each user visits, such as cafés, workplaces, or shops. When utilizing the service GPS and/or network, location data is extracted to provide the user with a list of nearby locations, and points are awarded for every check-in. This point-scoring feature is but one aspect of a key dimension of Foursquare—its game-like features, which include earning badges for various milestones, particular patterns of locations checked into and visiting particular locations, and the interuser competition to become "mayor" of a location by visiting that location on the most days in the preceding two months. The other important aspect of Foursquare consists of its social features, that is, sharing one's location with friends within the service and the option of linking with Facebook or Twitter accounts to automatically post updates to those services.

The theoretical framework that gives shape and direction to the analysis here is the political economy of the media approach. Historically, at least, much of this work has been concerned with broadcast media in one form or another. More recently, however, there has been considerable effort to explore the political economy of "new media"—or what Dwayne Winseck calls "the multiple economies of network media."[2] One productive account of what political economy can bring to the study of new or networked media was developed by Robin Mansell when she called for "a revitalization of research on new media in the tradition of political economy."[3] Drawing on the work of earlier political economy theorists and critics, Mansell argues that "any political economy of new media must be concerned with symbolic form, meaning and action as it is with structures of power and institutions"[4] and with the ways that "power is structured and differentiated, where it comes from and how it is renewed."[5] With its core check-in mechanic and the addition of a geolocational layer to the Internet in the form of the proprietary database that results, Foursquare is thus a rich case study for a political economic analysis in the mode Mansell argues for.

Navigating Foursquare's movement toward a sustainable economic basis is a key issue that a political economic analysis can address. Mansell highlights the ways in which scarcity remains an important dynamic despite "the abundance and variety of new media products and services" often emphasized by new media critics.[6] More precisely, she writes, "The production and consumption of new media in their commodity form means that scarcity has to be created by, for example, [the] creation and sale of audiences."[7] It is this production of artificial scarcity that underlies Foursquare's corporate strategy in developing a sustainable business model based on their propriety database of locations and, importantly, on users' patterns of interactions with those locations, particularly via data accrued through third-party access to its application programming interface (API).

A political economic approach also affords a way of understanding Foursquare's relation with and, significantly, reliance on its user base. Of particular importance is the emphasis, in Golding and Murdock's words, on the "concrete consequences" of patterns of power and ownership "for the work of making media goods."[8] While the dependence on UGC has become a central pillar of the cultural-economic logic of new media since its identification as a key principle of the Web 2.0 model,[9] Foursquare has sourced not only content from its users but also some of its key features and characteristics, such as the provision of discounts to mayors of locations, and has drawn heavily on user evangelism of the service.

In developing a political economic analysis of Foursquare, two particular issues present themselves. The first pertains to research resources. As José van Dijck and Thomas Poell point out, although "the underlying principles, tactics, and strategies" of social media platforms are generally apparent, the intricate connections between platforms and the internal business underpinning these connections and other economic arrangements are more difficult to chart.[10] Consequently, we draw extensively from the available trade press literature in developing this account of Foursquare's evolving engagement with location and related issues. This material provides continuity of data collection and makes possible an examination of the discursive strategies that are being constructed about and around location-based social media services by Foursquare and by industry commentators. The second issue relates to delimiting the scope of our analysis. Any attempt to construct a political economy of new media risks eliding key differences among various new media—or location-based—platforms, technologies, or services. Given this, we are highly conscious of N. Katherine Hayles' calls (made in another, very different context) for "medium-specific analysis."[11] In this chapter, we aim to develop, in Gerlitz and Helmond's words, "a platform critique that is sensitive to technical infrastructure while giving attention to the social and economic implications of the platform."[12] Thus, the approach adopted here is an exploration of a specific case concerning the history and evolving business model of Foursquare. The benefit of this approach is that it sheds light on quite specific cultural and corporate practices and economies in a way that potentially illuminates broader social media industry and consumer shifts.

FOURSQUARE: IN SEARCH OF A BUSINESS MODEL

In 2000, Dodgeball, the precursor to Foursquare, was developed by Dennis Crowley and Alex Rainert and purchased by Google in 2005. Crowley left Google when it became clear that Dodgeball was not going to receive further engineering resources from Google. Dodgeball was eventually shelved by Google in 2009, replaced by Google Latitude, and rolled into its Google Places API. Determined to continue developing the Dodgeball concept, Crowley started Foursquare with Naveen Selvadurai in 2009 and was later joined by his Dodgeball cofounder, Alex Rainert, the following year.[13] Crowley chose to base Foursquare on the technology start-up cluster emerging in New York City, the so-called Silicon Alley,[14] which would seem to be a natural fit for the urban, pedestrian-focused locative media service his company developed.

By September 2013, Foursquare had attracted in excess of 40 million users[15] (jumping from 10 million in 2011)[16] and had surpassed 4.5 billion total check-ins[17] (up from 1 billion in 2011).[18] Going on 2011 statistics, there is an even 50/50 split between male and female users, and 50 percent are outside the United States.[19]It is also no longer solely an Anglophone service, with French, Italian, German, Spanish, and Japanese support added in February 2011 and Korean, Portuguese, Russian, and Thai support added in September 2011;[20] others have followed since, including Turkish support.

A common question raised in relation to technology start-up companies, as well as the emerging new sectors of the media and communication industries that they comprise, is the underlying business models by which such companies seek to raise revenue and eventually operate at a profit. This is a shared concern for investors and other business-oriented analysts trying to gauge the likely success of these new enterprises and their suitability as investment and/or acquisition targets; for media scholars interested in changing patterns of media economics, the significance of the new types of mediated practices and services these new companies support, and the production and consumption of media content; and for policy makers charged with formulating and applying the regulatory frameworks under which these companies will operate. In the midst of what some commentators are describing as the beginning of a new bubble in the technology sector—speculation fueled by Twitter's initial IPO valuation at US$14.2 billion[21] and online storage company Dropbox's possible valuation at US$8 billion[22]—with neither company turning a profit—attention to the potential sustainability of these new ventures is particularly pertinent. This is especially so if we are to move beyond the techno boosterism that all too often accompanies the emergence of new types of media businesses and the associated sweeping claims of how they will revolutionize our everyday practices to arrive at a more considered analysis and evaluation of the significance of location-based social media services.

As the leading, established player in the field, Foursquare presents a valuable case study for considering the logic and assumptions informing the development of mobile location services—in particular, the approaches and strategies formulated to build a viable and sustainable business model. This section will discuss several key themes relating to Foursquare's approach to this challenge, specifically the importance of acquiring and maintaining a critical mass of users, experimentation with differing revenue models, the open sourcing or outsourcing of both business and software platform development, and the potential for partnerships and/or acquisition by other technology sector companies.

If we were seeking a general description of Foursquare's overall approach to date, which was both broad yet succinct, one effective candidate would be by way of comparison with Twitter. Much like Twitter (among others),[23] Foursquare has focused on acquiring a critical mass of users first, a strategy we could describe, to borrow the classic line from the film *Field of Dreams*, as the build-it-and-they-will-come approach. Indeed, as Foursquare cofounder and CEO Dennis Crowley stated during a question-and-answer session at the 140 Characters Conference in New York in June 2011, "the point isn't to become profitable right now, the point is to grow as quickly as possible, to really push the boundaries of what you can do with location based services on mobile devices."[24] There are some definite advantages to this approach, such as being able to try out different revenue models to see how they fit the platform and its users, as well as the flexibility to develop the platform in line with changing user expectations and competitive pressures. However, it also represents a certain leap of faith that the platform can attract enough users, media attention, and, most importantly of all, sources of venture capital to operate until the platform matures and develops a sustainable base and revenue model.

Foursquare has been moderately successful, especially early on, in attracting capital investment, raising a combined total of US$70 million in Series B and C venture capital rounds[25] and maintaining a global user base of over 40 million. Furthermore, Crowley's persona, online lifestyle, and promotional efforts through media interviews and presentations and appearances at various conferences have seen him emerge as a minor celebrity CEO who attracts individual coverage from the online technology press in addition to that which his company receives for its announcements and milestones. For example, his live blogging during the Hurricane Irene scare in New York was reported on by the *Business Insider* website;[26] his appearance with Foursquare cofounder Selvadurai in an advertisement for Gap clothing has been taken as an indication of his arrival as one of the leading members of a new generation of "tech world rockstars";[27] and his recent marriage to long-term girlfriend Chelsa Skees was covered by the *New York Times*.[28] More broadly, profiles of his life story and lifestyle have appeared in various publications, providing him with further opportunity to "tell the story," as he puts it, of Foursquare.[29]

Crowley's public profile, along with the enthusiastic and charismatic public persona he projects and his own personal history within the technology sector,[30] are all-important in providing him the opportunity to be an evangelist for the "story" or "vision" of Foursquare. For example, at an interview at 2011's South by South West Interactive, Crowley stated: "We always talk about making cities easier to use, making neighborhoods easier to use. We really want to build these tools that bridge the gap between what you experience online and how you take that offline."[31] Within this context of the "vision" of Foursquare, the game-like elements are intended to encourage users to explore their city, though the mechanics of the "mayor" system would suggest that this goal is not always achieved.[32]

Although Foursquare is yet to establish a viable business model, this does not mean that they are completely without revenue; the company has experimented with various income streams. One revenue source has been marketing partnerships and advertising deals with established, significant brands in a series of trial programs. In part, the company's strong focus on its vision and on developing the quality of its service to users has played a role in attracting this commercial interest,[33] and even though the financial returns for Foursquare to date have been modest in large part, the media exposure gained by the company, particularly in relation to some of the earlier deals they have secured, along with the symbolic vote of confidence from major mainstream brands, have arguably been far more valuable in raising the company's profile. For example, although a partnership in early

2010 with the U.S. cable television channel Bravo did not provide much direct revenue, the exposure helped introduce Foursquare to a mainstream audience outside their then mostly early adopter user base.[34] Given that Foursquare faces competition from much larger players in both the technology and social networking spheres, such as Google and Facebook,[35] the value of such exposure and the positive perceptions of its status as a serious business from its association with multinational brands is clearly part of a longer-term strategy with potential benefits that would outweigh considerations of short-term sources of revenue.

Even though Foursquare may follow countless other technology startups and be acquired by one of the larger established technology companies, much like their competitor Gowalla, which was acquired by Facebook at the end of 2011,[36] Crowley's previous experience with Google's acquisition of Dodgeball suggests that such an outcome is, for the moment, unlikely. There has been speculation by some within the trade press that Foursquare should merge with or be acquired by another company,[37] including the daily deals service Groupon, in order to provide it with a stable income and business model.[38] Despite no such acquisition having yet occurred, Foursquare has entered into partnerships both with Groupon and with other daily deals services companies to provide users with information on nearby deals through the Foursquare app.[39] Such partnerships that complement and augment the basic functionality of Foursquare are likely to continue being important for Foursquare as it attempts to build a stable revenue model for its service.

OPEN-SOURCED OR OUTSOURCED? FOURSQUARE' S API

An important aspect of Foursquare's ongoing efforts to build a stable revenue base is their decision, widely commented on in the trade press, to develop and release an open API, which has allowed third-party developers to add various functions and to develop unexpected uses of the platform, much like iPhone apps have added functionality and value to the iOS platform.

The provision of API tools has become common practice and part of the accepted wisdom for developing Internet services into platforms, although the recent tendency has been toward restricting third-party access to them. In the case of Foursquare, not only has it bucked recent trends by maintaining an open API,[40] the style of crowd-sourced development it permits has extended beyond the technical sphere. As Crowley has remarked, "We've been crowd sourcing pretty much everything."[41] Crowley has even remarked in interviews about the importance of user evangelism and of encouraging business owners to make use of the service after the "mayor" of a location makes the owner aware of Foursquare through asking about their reward; Crowley refers to this as "crowd sourcing the sales force."[42]

Foursquare's decision to provide API tools early in the development of the platform reaped almost immediate results, with an Android version of the mobile application being developed by hobbyist third-party developers to sit alongside the in-house-developed iOS application, with Crowley going as far as saying that the Android version was "better than the iPhone app we built ourselves."[43] As Jeff Watson notes, the creators of such additional applications and other additions and extensions to a platform like Foursquare are generally technically competent early adopters or lead users, who benefit from their unremunerated contribution by influencing the development of the platform toward their own desired uses.[44] In addition, many applications that are developed on the Foursquare platform by making use of its API are developed by other tech start-up companies, with some drawing on the APIs of other platforms, such as Google Maps, Facebook, and Twitter, to create extensions between these more established platforms.

One relatively straightforward type of addition is organizing or accessing the information available within the Foursquare platform in particular ways. For example, FourGraph creates an infographic detailing information regarding check-in locations, other statistics, and earned badges, and FourSquare Lists allows users to create lists of particular Foursquare venues to share with friends.[45] Other additions, such as (the now defunct) Sonar and Agora, made use of the API of various platforms to suggest people at a location that a user has checked into who might be appealing due to shared interests; When Should I Visit uses check-in data to determine the busiest and quietest times at cultural attractions in London.[46] Another common type of addition makes use of the API to further develop the game-like elements of Foursquare, such as World of Fourcraft, where users ally themselves with particular teams based on the boroughs of New York, which then compete for ownership of particular neighborhoods based on the number of check-ins for each team,[47] or ARstreets, an augmented reality graffiti iPhone app that shares the location of its users' virtual tags through Foursquare and other social networking platforms.[48]

Beyond the technical API, crowd-sourcing dynamics have played a key role in the development of the nature of the services offered through Foursquare. For example, the provision of discounts by venues to the mayor of a particular location or even to users who have checked into that location emerged not from Foursquare but rather from the venues themselves.[49] Noticing this trend, Foursquare then incorporated such possibilities into what would become its venue API. This particular example demonstrates well the advantage of the flexibility Foursquare has due to its lack of a specific business model; they are able to adapt to and incorporate the ways the actual end users of their service are using the service and thus develop organically and in some accordance with the desires of the user base.

The end users' key role in developing the usefulness and value of the service to other users has arguably been in populating cities with checked-in locations, not simply because this not only makes the process of checking-in to already established venues easier for subsequent users but also provides various kinds of information and tips, such as locations of local significance.[50] Beyond the adding of functional value for other users, there is also the other key output of users' interaction with Foursquare, as with other social networking sites. The accumulation of large amounts of data concerning people's movement patterns through urban space and their patterns of consumption has substantial financial value for advertisers and marketers and provides the opportunity for Foursquare to develop new features that make use of that data, for example, suggesting itineraries to users visiting another city.

Crucially, the pool of location data accumulated within Foursquare's places database is further enriched through cross-platform partnerships with other social networking services providing geolocation functionality via Foursquare's API. At present, such deals have been struck with Google's Vine and Waze, Facebook's Instagram, Yahoo!'s Flickr, as well as Uber, Evernote, Path, and Pinterest.[51] Following Schiller, it can be seen that Foursquare has actively sought to make itself an interstitial part of social networking services' "existing forcefield of institutional structures and functions."[52]

Recent interviews given by various members of Foursquare's executive team suggest that the company has begun to turn its attention more seriously toward developing suitable revenue models to leverage their user base (and places database)—while importantly not alienating the early adopters who have taken up its platform with enthusiasm—by moving toward an overt commercialization of the platform. There seems to be a shift in focus from the basic mechanics and incentives relating to the check-in process toward the provision of various kinds of content, such as the sorts of information and tips provided by users just mentioned, as well as that provided by partner organizations. As Crowley has stated, "for a lot of people it's been games about check-ins, and that's great, but like we're trying to do something much larger, this big data play about things that going on in the real world and connecting people to things in the real world."[53]

FROM BADGES TO "LOCATION AND DISCOVERY"

This strategic shift in direction is generally regarded as a response to persistent questions from industry analysts that the company has faced "about the long-term sustainability of its business."[54] Due to reported slowing in user growth[55] and underwhelming revenue generation to date, investors are said to be reluctant to give the company additional injections of capital investment.[56] Indeed, the fact that Foursquare's last round of investment (US$41 million) was a refinancing of debt, rather than an injection of fresh equity, is significant in this context for at least two reasons. It could indicate that investors are reluctant to commit fresh funds to the start-up. However, it is useful for Foursquare in that it enables them to delay a fresh valuation of the business which, if not improved on the 2011 valuation of US$600 million, could potentially hurt their prospects for ongoing growth or a successful exit.

These factors have prompted Foursquare to rethink its corporate strategy and business model. The path it appears to be taking is not new: the monetization of user traffic data. Elizabeth van Couvering points out that search and social media services tend to be characterized by media platforms that facilitate exchange between producers and audiences and that operate across "complex content pools that are large in size, extremely varied in terms of producers, and frequently refreshed."[57] Due to the complexity of the "content pool," the platform thus "becomes the central way to mediate connections between audiences and producers."[58] Significantly, "if the content pool is the network," then "audience traffic, enabled through the platform, are the connections within the network."[59] It is these "connections" that form the "core, saleable asset" for the owners of the platform.[60]

For Foursquare, the ability to parse geolocational and user check-in data is "key to its monetary strategy."[61] Foursquare's Dennis Crowley notes, "We are starting to get really good at figuring out what the context is" in the geodata accrued through check-ins[62] and using this check-in data as the basis for a search–and-recommendation platform. In short, the value of check-ins for Foursquare is the accumulation of large amounts of data concerning people's movement through urban space and their patterns of consumption. Significant both for Foursquare's monetization strategy and for a political economic analysis is the scarcity of this data. Thus, this information holds clear value for advertisers and marketers and provides the opportunity for Foursquare to develop new features that make use of that data. Three such features are discussed here.

The first of these is Explore, aimed at Foursquare users. In essence, it is a recommendations-and-ratings system—or, in the words of one commentator, an "interactive city guide"[63]—utilizing a series of metrics drawn from each user and his or her social network history (including tips, likes, dislikes, popularity, local expertise, and so on).[64] This information is then targeted to that user in the form of "recommendations for places you would probably like to visit based on your profile and check-in history."[65] Accompanying the Explore feature is an upgraded website, which is said to attract 1 million visits per day, with a prominent search box,[66] perhaps in recognition of the fact that a considerable amount of entertainment planning occurs at work or in the home. In late May 2013, Foursquare added what it calls "super-specific search" to Explore. This applies a range of filters to search results that combine common queries (such as price, opening hours, and so on), with additional information drawn from check-ins and user data. By September, restaurant menu search capabilities had also been added.[67] In the words of one tech commentator, "the end result isn't far off what Facebook is attempting to do with its ambitious Graph Search."[68] In short, in CEO Dennis Crowley's words, Foursquare is no longer about leader boards, badges, and points; it's about local search and discovery.

In addition, new features have been developed specifically for businesses. These fit squarely with Foursquare's plans to "get most of its future sales from software that helps merchants track the behavior of potential customers."[69] Even though, as noted, Foursquare already collects some revenue through strategic partnerships with competitors and a variety of companies,[70] these recent developments are quite different in that they want businesses to pay for help in analyzing the data generated through Foursquare's service by its users.[71] Foursquare's "first revenue-generating product,"[72] launched in 2012, was "promoted updates," advertising messages sent to users who are in the vicinity of a restaurant or other business. What distinguishes this service is that, rather than buying "advertising impressions," participating brands pay on a "cost-per-action model related to how consumers interact with the updates."[73] The second business feature is the Foursquare for Business app.[74] Launched in early 2013, it allows businesses to offer "digital punchcard" deals when users check in, in addition to sending messages to regulars.[75] And by October 2013, they opened up Foursquare Ads to all small businesses around the world.[76]

CONCLUSIONS

In this chapter, we have built on existing analyses of the political economies of new media by tracing Foursquare's still evolving business model and by documenting the company's dramatic strategic change in a direction where it has systematically downplayed its game play functionality[77] and played up the urban exploration aspects of its service. These moves work to deliberately reposition the company as embracing what has recently been termed the "like economy."[78] By relying heavily on both their interconnections with and between more established competitors to enrich their database, as well as their core group of super users to provide content and some of the central features of their service, Foursquare has followed and continues to follow a precarious path.

The nimbleness that Foursquare has displayed in making such a dramatic operational shift is impressive and clearly very difficult for larger corporate players to replicate. This is a key feature of technology start-ups in the early phases of their development and maturation as companies. In political economic terms, it is, however, a particularly high-risk maneuver for Foursquare for at least three reasons. First, it is potentially risky in terms of its possible impacts on its core user base. The strategy focuses on Foursquare's so-called super users, the "contingent of users and merchants who are seriously engaged by its platform." Once the company has a "dense, engaged, revenue-generating core of users," its aim is to expand from this base.[79] The trick for Foursquare will be in managing commercial and new user growth without ostracizing its core constituency of high-end users—those initially drawn to its game play elements.

Second, it is also a risky move in financial terms. Over its relatively short life, Foursquare has managed to attract investment funding totaling

US$112 million. But venture capital investment is by no means an endless stream, and the individuals and companies who have backed the company will want to see a return on their investment. Unless Foursquare can continue to grow its user base and/or its profit margins, the likely outcome will be a loss-making (and potentially embarrassing) exit.

Third, the new path Foursquare has embarked on is also risky insofar as it now places Foursquare in increasing competition with a number of larger, financially well-heeled competitors, including Google, Facebook, and Yelp, as well as Twitter. However, unlike these other platforms, Foursquare has a user base already acclimatized to location sharing and an emerging ability to interpret and leverage this data. Foursquare is banking on the granulation of this location and recommendation information to maintain its place in the social media ecosystem.[80]

Yet one of the more perplexing features of this location services ecosystem is that, by allowing Instagram, Vine, and Waze to access location data from Foursquare's API, Facebook and Google are actively contributing to the further enrichment of the "location 'underlayer' database for the Internet" that Foursquare is building.[81] It is this database, populated more extensively from these sources than from Foursquare's own users, that could in fact prove to be Foursquare's greatest asset and potential trump card. So long as this data stream persists (which is by no means assured), Foursquare's places database will become further enriched, fueling Crowley's desire for Foursquare to become *the* "location layer of the Internet," as well as pushing the firm closer to profitability or, at very least, an acceptably higher exit valuation.

ACKNOWLEDGMENTS

This chapter is an output from a program of research under an Australian Research Council Early Career Research Award—DE120102114, "The Cultural Economy of Locative Media," funded 2012–2014. The authors gratefully acknowledge the ARC's financial support; they would also like to thank Gerard Goggin and Jordan Frith for their feedback on this chapter, as well as Emily van der Nagel for her research assistance.

NOTES

1. Dan Schiller, "Deep Impact: The Web and the Changing Media Economy," *Info* 1, no. 1 (February 1999): 35.
2. Dwayne Winseck, "The Political Economies of Media and the Transformation of the Global Media Industries," in *The Political Economies of Media: The Transformation of the Global Media Industries*, eds. Dwayne Winseck and Dal Yong Jin (London: Bloomsbury, 2011), 7.
3. Robin Mansell, "Political Economy, Power and New Media," *New Media & Society* 6, no. 1 (2004): 96.

4. Mansell, "Political Economy," 98.
5. Mansell, "Political Economy," 99.
6. Mansell, "Political Economy," 97.
7. Mansell, "Political Economy," 98.
8. Peter Golding and Graham Murdock, "Culture, Communications and Political Economy," in *Mass Media and Society*, 3rd ed., eds. James Curran and Michael Gurevitch (London: Arnold, 2000), 84.
9. For example, see Tim O'Reilly, "What Is Web 2.0: Design Practice and Business Models for the Next Generation of Software," *Communications and Strategies* 61, no. 1 (2007): 17–37.
10. José van Dijck and Thomas Poell, "Understanding Social Media Logic," *Media and Communication* 1, no. 1 (2013): 2.
11. N. Katherine Hayles, "Print Is Flat, Code Is Deep: The Importance of Media-Specific Analysis," *Poetics Today* 25, no. 1: 67–90.
12. Carolin Gerlitz and Anne Helmond, "The Like Economy: Social Buttons and the Data-Intensive Web," *New Media & Society* [online first] (2013), accessed December 6, 2013, doi: 10.1177/1461444812472322; see also Tarleton Gillespie, "The Politics of 'Platforms,'" *New Media & Society* 12, no. 3 (2010): 347–364.
13. Leena Rao, "The Gang Is Back Together," *TechCrunch* (April 15, 2010), accessed November 27, 2013, http://techcrunch.com/2010/04/15/the-gang-is-back-together-dodgeball-co-founder-joins-foursquare-as-product-chief/.
14. For detailed discussion, see Gina Neff, *Venture Labor: Work and the Burden of Risk in Innovative Industries* (Cambridge, MA: MIT Press, 2012).
15. "About Foursquare," *Foursquare* (September 2013), https://foursquare.com/about.
16. Pascal-Emmanuel Gobry, "Foursquare Gets 3 Million Check-Ins Per Day, Signed Up 500,000 Merchants," *Business Insider* (August 2, 2011), accessed November 27, 2013, www.businessinsider.com/foursquare-dennis-crowley-interview-2011–8.
17. "About Foursquare," *Foursquare* (September 2013), accessed November 27, 2013, https://foursquare.com/about.
18. Alyson Shontell, "Foursquare Passes 1,000,000,000 Check-Ins," *Business Insider* (September 20, 2011), accessed November 27, 2013, http://articles.businessinsider.com/2011–09–20/tech/30179114_1_foursquare-check-ins-app.
19. Alexa Tsotsis, "Foursquare Now Officially at 10 Million Users," *TechCrunch* (June 20, 2011), accessed November 27, 2013, http://techcrunch.com/2011/06/20/foursquare-now-officially-at-10-million-users/ .
20. Rich Harris, "Foursquare Adds Five More Languages to the Mix," *ZDNet* (September 8, 2011), www.zdnet.com/blog/feeds/foursquare-adds-five-more-languages-to-the-mix/4121.
21. Katherine Rushton, "Twitter Worth $14.2bn as Social Network Prices IPO at $26 a Share," *The Telegraph* (November 6, 2013), accessed November 27, 2013, www.telegraph.co.uk/technology/twitter/10431560/Twitter-worth-14.2bn-as-social-network-prices-IPO-at-26-a-share.html.
22. Douglas MacMillan and Spencer E. Ante, "Dropbox Seeks Funding Round at $8 Billion Valuation," *Wall Street Journal* (November 18, 2013), accessed November 27, 2013, http://online.wsj.com/news/articles/SB10001424052702 303985504579206763922615986.
23. Matthew Yglesias, "The Prophet of No Profit: How Jeff Bezos Won the Faith of Wall Street," *Slate* (January 30, 2013), accessed February 12, 2014, www.slate.com/articles/business/moneybox/2014/01/amazon_earnings_how_jeff_bezos_gets_investors_to_believe_in_him.html.

24. 140 Talks, "OA with Dennis Crowley," *#140Conf NYC 2011* (December 20, 2011), accessed November 27, 2013, www.youtube.com/watch?v=K-CdwtX0ci0.
25. Sarah Lacey, "Foursquare Closes $50M at a $600M Valuation," *TechCrunch* (June 24, 2011), accessed November 27, 2013, http://techcrunch.com/2011/06/24/foursquare-closes-50m-at-a-600m-valuation/.
26. Alyson Shontell, "Dennis Crowley's Great Hurricane Irene Adventure," *Business Insider* (August 29, 2011), accessed November 27, 2013, www.businessinsider.com/hurricane-irene-new-york-dennis-crowley.
27. Noah Davis, "Tech CEOs Are the New All-Star Pitchmen," *Business Insider* (September 19, 2011), accessed November 27, 2013, www.businessinsider.com/tech-ceos-are-the-new-all-star-pitchman-2011–9.
28. Margaux Laskey, "'Girl Version' of Him, 'Boy Version' of Her," *New York Times* (October 20, 2013), accessed November 27, 2013, www.nytimes.com/2013/10/20/fashion/weddings/girl-version-of-him-boy-version-of-her.html?_r=3&.
29. See, most recently, Austin Carr, "The Lost Boy," *Fast Company* 178 (September 2013): 84–90, 106–107.
30. For example, see Edmund Lee, "Is Dennis Crowley the Pied Piper of Silicon Alley?" *AdAge Digital* (August 8, 2011), accessed November 27, 2013, http://adage.com/article/digital/foursquare-s-dennis-crowley-silicon-alley-s-pied-piper/229052/; Nicholas Carlson, "Dennis Crawley Parties Non-Stop, and It Is the Reason He's Such a Brilliant Leader at Foursquare," *Business Insider* (August 8, 2011), accessed November 27, 2013, www.businessinsider.com/dennis-crowley-parties-non-stop-and-it-is-the-reason-hes-such-a-brilliant-leader-at-foursquare-2011–8.
31. Dennis Crowley, "Dennis Crowley on the Future of Foursquare," *Mashable* (March 16, 2011), accessed November 27, 2013, www.youtube.com/watch?v=tQOgHN1VTKk.
32. For critiques of Foursquare along these lines, see Alison Gazzard, "Location, Location, Location: Collecting Space and Place in Mobile Media," *Convergence* 17, no. 4 (2011): 405–417; David Phillips, "Identity and Surveillance Play in Hybrid Space," in *Online Territories: Globalization, Mediated Practice and Social Space*, eds. Miyase Christensen, André Jansson, and Christian Christensen (New York: Peter Lang, 2011), 171–184.
33. Lee, "Is Dennis Crowley the Pied Piper of Silicon Alley?"
34. Kunur Patel, "Foursquare Plots Its Business Model," *Business Insider* (February 4, 2010), accessed November 27, 2013, www.businessinsider.com/foursquare-plots-its-business-model-2010–2.
35. Interestingly, one industry commentator argues that Facebook's entry into locative media services with Facebook Places has helped rather than hindered Foursquare by exposing location check-in-based services to its vast user base. Dan Rowinski, "2Way Summit Preview: Who's Leading the Future of Location," *ReadWriteWeb* (June, 10, 2010), accessed November 27, 2013, www.businessinsider.com/foursquare-plots-its-business-model-2010–2.
36. According to reports, Facebook is not acquiring Gowalla's technology or user data;rather, the founders and key staff members will be joining Facebook as employees. Josh Constine, "Gowalla Confirms It Will Shut Down as Founders and Team Members Join Facebook," *TechCrunch* (December 5, 2011), accessed November 27, 2013, http://techcrunch.com/2011/12/05/gowalla-acqhire/.
37. Eric Markowitz, "Who Will Buy Foursquare?" *Inc.* (August 30, 2013), accessed November 27, 2013, www.inc.com/eric-markowitz/who-will-buy-foursquare.html.

38. For examples, see Ben Parr, "Should Groupon Acquire Foursquare," *Mashable* (May 24, 2011), accessed November 27, 2013, http://mashable.com/2011/05/23/groupon-foursquare/; Nicholas Carlson, "Groupon to Buy Foursquare (If We Can Negotiate the Deal over Twitter," *Business Insider* (May 17, 2010), accessed November 27, 2013, www.businessinsider.com/groupon-to-buy-foursquare-if-we-can-negotiate-the-deal-over-twitter-2010–5.
39. Jennifer Van Grove, "Foursquare & Groupon Hook Up for Real-Time Deals," *Mashable* (July 29, 2011), accessed November 27, 2013, http://mashable.com/2011/07/29/foursquare-groupon-partnership/.
40. John Sheehan, "APIs Are Dead, Long Live APIs," *The Next Web* (March 12, 2013), accessed November 27, 2013, http://thenextweb.com/dd/2013/03/12/apis-are-dead-long-live-apis/.
41. Zeitgeistminds, "Tim O'Reilly, Jerry Yang, Dennis Crowley, and Steve Cousins—US Zeitgeist 2010," *Zeitgeistminds* (October 1, 2010), accessed November 27, 2013, www.youtube.com/watch?v=yyCadszyeos.
42. Zeitgeistminds, "Tim O'Reilly, Jerry Yang, Dennis Crowley, and Steve Cousins—US Zeitgeist 2010."
43. Mac Slocum, "Foursquare Wants to be the Major of Location Apps," *O'Reilly Radar* (March 1, 2010), accessed November 27, 2013, http://radar.oreilly.com/2010/03/foursquare-location-apps.html.
44. Jeff Watson, "Fandom Squared: Web 2.0 and Fannish Production," *Transformative Works and Cultures*, no. 5 (2010), http://journal.transformativeworks.org/index.php/twc/article/viewArticle/218/183.
45. Chris Thompson, "Best of the Foursquare Hack Day," *About Foursquare* (February 21, 2011), http://aboutfoursquare.com/best-of-the-hack-day/.
46. Jennifer Van Grove, "6 New Apps for Getting More Out of Foursquare," *Mashable* (June 5, 2011), accessed November 27, 2013, http://mashable.com/2011/06/04/new-foursquare-apps/.
47. Sarah Kessler, "Clever Foursquare Hack Turns New York City into a Giant Game of Risk," *Mashable* (June 29, 2011), accessed November 27, 2013, http://mashable.com/2011/06/29/world-of-fourcraft/.
48. Kat Hannaford, "Augmented Reality Graffiti Hits Foursquare in ARStreets iPhone App," *Gizmodo Australia* (October 25, 2010), accessed November 27, 2013, www.gizmodo.com.au/2010/10/augmented-reality-graffiti-hits-foursquare-in-arstreets-iphone-app/.
49. Zeitgeistminds, "Tim O'Reilly, Jerry Yang, Dennis Crowley, and Steve Cousins—US Zeitgeist 2010."
50. Zeitgeistminds, "Tim O'Reilly, Jerry Yang, Dennis Crowley, and Steve Cousins—US Zeitgeist 2010."
51. Jim Edwards, "Facebook is Holding an Axe over the Neck of Foursquare," *Business Insider Australia* (October 20, 2013), accessed November 27, 2013, www.businessinsider.com.au/facebook-is-holding-an-ax-over-the-neck-of-foursquare-2013–10; Anil Dash, "On Location with Foursquare," *Anil Dash: A Blog About Making Culture* (August 6, 2013), accessed November 27, 2013, http://dashes.com/anil/2013/08/on-location-with-foursquare.html; Michael Calore, "How Foursquare Is Forcing Social Networks to Check In or Check Out," *Wired* (March 12, 2013), accessed November 27, 2013, www.wired.com/underwire/2013/03/location-apps-social-media/; "Introducing Foursquare-Powered Place Pins on Pinterest," *Foursquare* (November 22, 2013), accessed November 27, 2013, http://foursquare.tumblr.com/post/67672365878/introducing-foursquare-powered-place-pins-on-pinterest.
52. Schiller, "Deep Impact," 35.
53. Dennis Crowley, "Dennis Crowly on the Future of Foursquare."

54. Mike Isaac, "Foursquare's New App Is Open for Business," *All Things D* (January 29, 2013), accessed November 27, 2013, http://allthingsd.com/20130129/foursquares-new-app-is-open-for-business/.
55. Benedict Evans, "Foursquare Traction," *Benedict Evans* (October 9, 2013), accessed November 27, 2013, http://ben-evans.com/benedictevans/2013/10/9/foursquare-traction.
56. Seth Fiegerman, "Foursquare Will Make Just $2 Million in Revenues This Year," *Mashable.com* (November 22, 2012), accessed November 27, 2013, http://mashable.com/2012/11/21/foursquare-revenue-2-million-2012/.
57. Elizabeth van Couvering, "Navigational Media: The Political Economy of Online Traffic," in *The Political Economies of the Media: The Transformation of the Global Media Industries*, eds. Dwayne Winseck and Dal Yong Jin (London: Bloomsbury, 2011), 198.
58. van Couvering, "Navigational Media," 198.
59. van Couvering, "Navigational Media," 198.
60. van Couvering, "Navigational Media," 198.
61. David Goldman, "Foursquare CEO: 'Not Just Check-Ins and Badges,'" *CNNMoney* (February 29, 2012), accessed November 27, 2013, http://money.cnn.com/2012/02/29/technology/foursquare_ceo/index.htm.
62. Dennis Crowley, quoted in Goldman, "Foursquare CEO."
63. Pascal-Emmanuel Gobry, "Foursquare's Revenue Model Sharpens into Focus," *Business Insider* (May 9, 2012), accessed November 27, 2013, http://articles.businessinsider.com/2012–05–09/research/31634849_1_foursquare-business-model-yelp.
64. Dara Kerr, "Foursquare Launches Rating System, Competes with Yelp," *CNet.com* (November 5, 2012), accessed November 27, 2013, http://news.cnet.com/8301–1023_3–57545517–93/foursquare-launches-rating-system-competes-with-yelp/.
65. Goldman, "Foursquare CEO."
66. Kerr, "Foursquare Launches Rating System."
67. Greg Sterling, "Foursquare Adds Menu Items to Search Capability," *Search Engine Land* (September 9, 2013), accessed November 27, 2013, http://searchengineland.com/foursquare-adds-menu-items-to-search-capability-171505.
68. Chris Welch, "Foursquare Rolls Out 'Super-Specific' Filters to Help You Search Out the Perfect Destination," *The Verge* (May 22, 2013), accessed November 27, 2013, www.theverge.com/2013/5/22/4355862/foursquare-rolls-out-super-specific-natural-language-search.
69. Dennis Crowley quoted in Emily Chang and Douglas MacMillan, "Foursquare Says Merchant Services Will Provide Bulk of Revenue," *Bloomberg.com* (August 2, 2011), accessed November 27, 2013, www.bloomberg.com/news/2011–08–02/foursquare-ceo-says-merchant-services-will-provide-bulk-of-startup-s-sales.html.
70. Chang and MacMillan, "Foursquare Says Merchant Services;" Jennifer Van Grove, "Foursquare Checks in to More Revenue with Credit Card Specials," *CNet News* (February 25, 2013), accessed November 27, 2013, http://news.cnet.com/8301–1023_3–57571202–93/foursquare-checks-in-to-more-revenue-with-credit-card-specials/.
71. Chang and MacMillan, "Foursquare Says Merchant Services."
72. Seth Fiegerman, "Foursquare Will Make Just $2 million in Revenues This Year."
73. Anne Marie Kelly, "Three Reasons Why Foursquare's New Advertising Model Might Work," *Forbes* (August 22, 2012), accessed November 27, 2013, www.forbes.com/sites/annemariekelly/2012/08/22/three-reasons-why-foursquares-new-advertising-model-might-work/.

74. Foursquare Blog, "Manage a Business on Foursquare? Download Our New App to Easily Connect with Customers Right from Your Phone," *Foursquare Blog* (January 29, 2013), accessed November 27, 2013, http://blog.foursquare.com/2013/01/29/manage-a-business-on-foursquare-download-our-new-app-to-easily-connect-with-customers-right-from-your-phone/; Mike Isaac, "Foursquare's New App Is Open for Business."
75. Mike Isaac, "With New Merchant Local Updates Tool, Foursquare is Getting Serious About Its Business," *All Things D* (July 18, 2012), accessed November 27, 2013, http://allthingsd.com/20120718/with-new-merchant-local-updates-tool-foursquare-is-getting-serious-about-its-business/.
76. "Big News: Today We're Opening Up Foursquare Ads to All Small Businesses Around the World," *Foursquare Blog* (October 14, 2013), accessed November 27, 2013, http://blog.foursquare.com/2013/10/14/big-news-today-were-opening-up-foursquare-ads-to-all-small-businesses-around-the-world/.
77. Mike Isaac, "CEO Dennis Crowley on Foursquare's Biggest Mistake," *All Things D* (March 11, 2013), http://allthingsd.com/20130311/ceo-dennis-crowley-on-foursquares-biggest-mistake/.
78. Gerlitz and Helmond, "The Like Economy."
79. Gobry, "Foursquare's Revenue Model."
80. Calore, "How Foursquare Is Forcing."
81. Edwards, "Facebook Is Holding an Axe over the Neck of Foursquare."

14 Becoming Drones
Smartphone Probes and Distributed Sensing

Mark Andrejevic

> The drone is the most powerful eruption and the most beguiling expression of the transnational vortex. The reason it has become a pop-cultural phenomenon and an object of fascination and study for people in many different sectors is that it is an incandescent reflection, the most extreme expression of who we are and what we've become generally.[1]

Shortly after a former defense contractor and intelligence analyst Edward Snowden revealed to the world that the U.S. National Security Agency was collecting data about hundreds of millions of phone calls daily, telecommunications giant AT&T announced it was considering marketing data about its users.[2] If the data was so useful, why not find new markets for it, especially considering the fact that other telecommunication companies like Verizon (which was at the center of the NSA revelations) had already been selling user data for some time?[3] In both instances, what was at issue was so-called metadata—not the actual content of conversations but the detailed contextual information generated by mobile and smartphones, including details of time and location—especially location—and what this might reveal about users. 'In the wake of such revelations, the implied assumption that contextual information was not as invasive or intrusive as monitoring content—started to unravel. Former Sun Microsystems engineer Susan Landau, for example, confided to *The New Yorker* magazine that the public "doesn't understand" that metadata is "much more intrusive than content."[4] Examples multiplied across the journalistic landscape of the innovative and revelatory uses to which contextual information might be put: "A person who knows all of another's travels can deduce whether he is a weekly churchgoer, a heavy drinker, a regular at the gym, an unfaithful husband, an outpatient receiving medical treatment, an associate of particular individuals or political groups . . . and not just one such fact about a person, but all such facts."[5] Landau's examples were cited in several media outlets, including the observation that very specific conclusions about protected categories of personal information could be gleaned from metadata.[6] The revelatory role of location played a central and crucial role in such accounts—not just *directly*, in terms of the ability to capture data about the time–space paths of users, but also *indirectly*, insofar as the comprehensiveness of data capture is related to the mobility and the addressability of portable interactive devices. The goal of this chapter, then, is to provide a model for thinking about the combination of these various aspects of mediated locational tracking: the combination of dispersed, portable devices and centralized forms of mobile data capture, with new forms of location-based targeting and decision making.

Shortly after the discussion of the power of metadata prompted by the NSA revelations, a similar but connected set of observations emerged in the AT&T coverage. The company reported, for example, that it would be able to sell TV networks' data about its customers' TV viewing behavior (derived from its pay TV services) with "other aggregate information we may have about our subscribers" (a huge category, including data acquired from other companies and services) in order to help producers "better understand the audiences that are viewing their programs."[7] These somewhat vague descriptions were rapidly coupled in the news coverage with some of the more vivid claims about the value of telecommunications metadata. Unsurprisingly, location was a key theme of these examples, despite the apparently immobile character of most television viewing. A Verizon executive, for example, described the ways in which his company could "provide insights on where a customer travels through the community" in order to, among other things, target location-based advertising, including outdoor advertising and targeted interactive offers: "it's amazing how you can reach people through their mobile handset."[8] Although the spokesperson specified that the user data would be mined on an aggregate level, he noted that information capture was granular and individualized: "We're doing this on a one-to-one basis even though we're marketing on aggregate anonymously, because we're able to just view everything they do."[9] He described, for example, how the company can locate subscribers attending a particular sporting event and "analyse what people are viewing on their handsets . . . we can tell if you're viewing ESPN or MLB [the Major League Baseball outlet], we can tell what social networking sites you're accessing, we pretty much can slice and dice it any way that our customers want."[10] And by customers, of course, he means not Verizon's end-user subscribers but the clients for the data the company has managed to harvest about users who are paying for, among other things, being tracked. Location is a key dimension of customization and targeting because it remains the definitive identifier of individual uniqueness: two unique individuals cannot occupy the same space, and, by the same token, an individual cannot simultaneously occupy multiple spaces: specification, targeting, customization, addressability all rely on this crucial dimension of location.

Uniting the two examples of state and corporate surveillance is the locative dimension: how to saturate the space of surveillance so as to be able to track and target. In both cases, the answer is to treat the mobile phone (or other networked, interactive device) as a drone-like probe. As in the case of many other media technologies, we tend to think of mobile phones and other portable interactive technologies in terms of the forms of communication and information we derive from them, and not so much in terms of how they "see" us. This is perhaps a habit conserved from the heyday of preinteractive mass media: we watch TVs and radio; they do not watch and listen back. We read newspapers and magazines; they do not read us. Of course, in the digital era, media consumption runs both ways: the more continuously we use our various networked devices, the more comprehensive an ongoing self-portrait we generate. It is tempting to repurpose Raymond William's famous observation about TV as a cultural form for the digital era. Digital data collection might be understood as enabling a new form of reverse flow, that is, the ability to capture the rhythms of the activities of our daily lives via the distributed, mobile, interactive probes carried around by the populace. As one media executive has put it, "What we do specifically is we understand what our customers' daily activity stream is."[11] This is the obverse of viewers immersed in the flow of television programming: producers immersed in the flow of the daily lives of viewers.

Tracking an "activity stream" necessarily requires mobile or distributed devices that can follow users throughout the course of the day. Thus location becomes a new "frontier" for commerce (spawning various neologisms including "m-commerce," "mociology," and "mocioeconomics"), and mobility (or what amounts to the same thing, distributed, ubiquitous sensing) becomes a crucial attribute of interactive technology.[12] The multiplication of networked interactive devices disperses always-on, mobile sensors throughout the populace. The movement of mobile phones, for example, can be used to monitor traffic flow through the city or, at the individual level, to determine a user's activities. (Is this someone who stays in the office all day, or goes out for a lunchtime run? How many strides a day does a particular user take? And so on.) Mobility lies at the heart of this perspective of consumer data as flow because portability makes it easier to distribute information-gathering applications throughout the populace and to link particular users with individual devices. Digitization also lies at the heart of such strategies because it allows for the ready and automated storage, capture, and processing of the huge amounts of data generated by the proliferation of devices and the expanding array of services they provide.

"DRONING" THE MEDIA

One of the reasons the figure of the drone has so rapidly captured the popular and media imagination is that, in addition to reviving what might be described as the ballistic imaginary once associated with technological gadgetry (in the *Popular Science* vision of personal jet packs and rocket ships), it encapsulates the emerging logic of portable, always-on, distributed, ubiquitous, and automated information capture. The promise of the drone as hyperefficient information technology is four-fold: (1) it extends and multiplies the reach of the senses; (2) it saturates the times and spaces in which sensing takes place (entire cities can be photographed 24 hours a day); (3) it automates the sense-making process; and (4) it automates response. In this regard, the figure of the drone unites ballistic and information technology: it is not simply a weaponized mobile camera (this is only one modality in which the logic of the drone manifests itself) but an indefinitely expandable probe that foregrounds the seemingly inevitable logic of algorithmic decision making. It is a telling fact, for example, that drones have become notoriously associated with the so-called signature strike—that is, an attack based on the detection of a particular (and publicly unspecified) profile of behavior that allegedly identifies insurgent or terrorist activity.[13] We might describe this as the emergence of a semiautomated attack algorithm, facilitated by the fact that the device can be used to simultaneously collect information that helps determine the appropriate data signature and to act on it.[14] The decision-making process is not (yet?) fully automated, but the algorithmic character of the decision-making process is summed up in the form of the signature as opposed to the "personality" strike. A signature strike takes place based not on personal identification or particular known individuals but on the probabilistic logic of the profile: the observed behavior has the signature of terrorist or insurgent activity. This logic is an increasingly familiar one in the realm of data mining generally—whether for the purposes of health care, surveillance, marketing, policing, or security. Identification takes a backseat to data analytics: one needn't know the name of a particular individual to target him or her, merely that he or she fits the target profile.

As a proliferation of recent examples has suggested, the drone is not necessarily an attack device: it can be used to dust crops, film sporting events, deliver humanitarian aid, monitor weather patterns, guide disaster response, and so on. Yet the image of the drone carries with it unmistakably ominous overtones that resonate with its military and surveillance history and uses. This chapter argues for a broader reading of the figure of the drone in which the logics of mobility, sensing, ubiquity, and automation coalesce. The drone comes to stand as a contemporary icon of the (inter)face of new forms of monitoring and surveillance—an exemplar of the always-on, networked, mobile, sensing device. The reference to the figure of the drone is thus, necessarily, infrastructural: it refers not just to the sensing device but to the architecture of information collection, transmission, analysis, and response within which it is the most visible, high-profile feature. Broadly construed, then, the figure of the drone is an icon of locative media: it relies on spatial dispersion in order to pin sensor data to time–space location.

Discussions of big data, data mining, and new forms of monitoring and surveillance often emphasize the figure of the database, the place where the data is stored, rather than that of the infrastructure that makes data collection possible. In part, this is because of the distributed and heterogeneous character of the various sensors that comprise the monitoring "assemblage,"[15] but in part it is because of what might be described as the turn away from infrastructure that has characterized the fascination with so-called immaterial forms of activity. This turn is echoed in the rhetoric of immateriality that characterizes discussions of the "cloud" (in "cloud computing") and cyberspace more generally.[16] The figure of the drone focuses attention back on the interface device that serves as mediator for both information collection and a certain type of automated action or response at a distance. Instead of the airy, light, unfettered figure of the cloud as the face of new forms of information capture, storage, and—yes—response (the goal of commercial cloud computing is, among other things, increasingly more precise targeting), this chapter proposes the somewhat more intrusive, tethered figure of the drone.

It is with this broader conception of the drone in mind that we might explore the ways in which the logic of the distributed, efficient, portable probe is generalized across the landscape of digital interactivity. The hallmarks of the drone as a material object are—like so many of the digital devices that have come to permeate the daily life of technologically saturated societies—its mobility and miniaturization, that is, its anticipated efficiency as a ubiquitous, always-on probe for generalized rather than targeted information collection. Among other connotations, the name "drone" conjures up the image of a continuous background presence, and much of the development of the drone as sensor- and munitions-carrying device hinges on the ability to keep it in the air as long as possible in order to establish an always-on presence that lends itself to both surveillance and the perpetual warfare associated with the so-called war on terror. This version of constant connectivity parallels the forms of always-on relationship monitoring that characterize the digital economy. As recent revelations regarding government use of commercially generated data indicate, these logics are two sides of the same coin.

The goal of reading data mining through the lens of the drone is, to put it somewhat differently, to foreground the alternative perspective on the digital media devices just described: that of probes that combine information collection, automated decision making, and response. In short, this chapter proposes viewing the smartphone (and, by extension, the laptop, tablet, and other such devices) as (in addition to other functions) drone/probe. It also explores some of the resulting insights, both for an understanding of locative media as emerging remote-sensing networks and for exploring the ways in which these incorporate the drone logics of automation and action and sensing at a distance. The following analysis draws on two case studies, one from the public sector and one from the private: the U.S. Department of Homeland Security's Cell-All program and MoodScope, a Microsoft initiative to use smartphones as mood monitors. It concludes with a discussion of what might be described as "drone theory:" the fascination with posthuman forms of information processing in an era of automated data collection and response.

IF YOU SMELL SOMETHING, SAY SOMETHING: THE SMART PHONE AS HANDHELD DRONE

The U.S. Department of Homeland Security (DHS) has, from its inception, relied on crowd-sourcing campaigns ("If you see something, say something") to mobilize the populace in the face of a threat that it frames as ubiquitous and perpetual: a terror strike that could take place virtually anywhere, along a growing variety of threat vectors. Its proposed Cell-All program automates the crowd-sourcing process by embedding sensors in smart phones to detect toxic chemicals (and eventually other threats) and report them directly to the authorities. As the DHS publicity puts it, "Just as antivirus software bides its time in the background and springs to life when it spies suspicious activity, so Cell-All regularly sniffs the surrounding air for certain volatile chemical compounds. When a threat is sensed, a virtual *ah-choo!* ensues."[17] If the threat is a "personal safety issue," that is, an immediate undetectable threat such as toxic levels of odorless carbon monoxide, the phone sounds an alert. However, "For catastrophes such as a sarin gas attack, details—including time, location, and the compound—are phoned home to an emergency operations center."[18] The goal is to piggyback on the proliferation of smartphones to create a massively distributed, mobile, sensor array that goes wherever people do, "casting a wider net than stationary sensors can."[19]

According to the DHS proposal, the system would operate on an opt-in basis with the injunction that participation in the system serves as a form of both personal protection and national defense. Unlike other forms of crowd-sourcing, once one opts into the sensor system, the response is automated; interactivity becomes passive: "anywhere a chemical threat breaks out—a mall, a bus, subway, or office—Cell-All will alert the authorities automatically. Detection, identification, and notification all take place in less than 60 seconds. Because the data are delivered digitally, Cell-All reduces the chance of human error."[20] The infrastructure for what Bill Gates once called friction-free capitalism redoubles itself in the figure of frictionless securitization—the friction, in both cases, provided by the users themselves: "Currently, if a person suspects that something is amiss, he *might* dial 9-1-1, though behavioral science tells us that it's easier to do nothing. If he does do something, it may be at a risk to his own life. And as is often the case when someone phones in an emergency, the caller may be frantic and difficult to understand."[21] In contrast, the automated response of distributed probes bypasses the various shortcomings of their human bearers, whose senses might not even be able to detect the threat permeating their environment.

All of which is not to discount the potential benefits of such a system but rather to point out the logic that links them to broader strategies for locative-media-enabled automated sensing, information processing, and response. The Cell-All project builds on and reflects an understanding of interactivity that characterizes the commercial use of distributed digital devices: they are not simply portable content-sharing and communication devices but also mobile probes that allow for new forms of centralized information collection and processing. The devices harness users' movements and in turn provide location-specific data for centralized response. In this context, the humans who carry them might be conceptualized as propulsion devices that allow the probes to circulate in populated spaces according to the rhythms of human activity: "casting a wider network than stationary sensors can."[22] Taken to the limit, the Cell-All network is meant to be coextensive with the communication network. As the DHS publicity puts it, the scheme, "envisions a chemical sensor in every cell phone in every pocket, purse, or belt holster. If it's not already the case, our smart phones may soon be smarter than we are."[23] That is to say, they might be more "aware" of their environment in ways that can be automatically registered elsewhere. This awareness is installed as an extra communication layer on top of the point-to-point communication associated with telephony and the one-to-many form of net casting enabled by mobile apps such as Twitter or Facebook. It inserts the monitoring capability into the communication manifold of networked, digital devices. As the technical director for the Homeland Security Advanced Research Projects Agency (HSARPA) put it, "What we're trying to do here is make chemical sensors part of the fabric of society."[24]

This colonization of the spaces of daily life, of course, is also the logic of the drone: not an event-driven, discrete form of monitoring but a continuous background presence, able to detect threat where and when it happens. The conceptual structure of this kind of distributed remote sensing is flexible and expandable, depending on the development of new sensing technology. The mobile phone, like the drone more generally, can be equipped with a growing array of sensors to enable a proliferating range of activities that go beyond national security narrowly construed to encompass health care, agriculture, commerce, and so on. As HSARPA's technical director put it, "There are a million and one uses for this."[25]

POPULACE AS PROBE: MOOD MONITORING

The current frontier of smartphone sensing is that of so-called affective computing: the attempt to gather information about users' emotional states based on interactively generated data. In this regard, we might describe mobile phones as always-on mood probes, circulating among the populace in order to harvest useful affective data for purposes ranging from marketing to academic research to therapy and health care. As one group of researchers put it, "Smartphones are ubiquitous, unobtrusive, and sensor-rich computing devices, carried by billions of users every day. More importantly, owners are likely to 'forget' their presence, allowing for the passive and effortless collection of data streams that capture user behavior."[26] The observation bears some similarity to discourses about the forms of "always-on" monitoring that characterize reality TV: the more comprehensive the monitoring, the more naturalized and thus more invisible it becomes. The goal of such applications is to integrate themselves into the fabric of daily life so that the information collection process becomes both omnipresent and invisible, like air.

This type of information collection already takes place in a range of registers, including, for example, location tracking via mobile phones, which serve as always-on location probes. Police have already used mobile phone data, for example, to catch thieves by placing them at the scene of the crime and reconstructing their movements in a subsequent car chase.[27] Mobile phones can be used not just to track aggregate traffic flow but also to monitor the movements of consumers through retail spaces.[28] As these forms of population monitoring become increasingly normalized, the next frontier of sensor-based data capture is the realm of mood, emotion, and even personality type.

Interest in the affective register is based not simply on its novelty but also on contemporary theories of influence and decision making associated with developments in neuroscience facilitated by new brain-imaging technologies.[29] Marketers, for example, are particularly interested in affective states and influences because of the impact these might have on consumption decisions. Moreover, because of the posited role of affective response in providing users with a "shortcut" through the welter of available information (and simultaneously offering marketers a way to cut through the clutter of competing ads), there is increasing interest in the possibility that mood might help play an important role in the forms of automated filtering that help consumers navigate a growing array of choices. The developers of, for example, recommendation algorithms that suggest movies or music for users are turning to the affective frontier of mood detection and monitoring. The proliferation of information generated by social media sites of one kind or another has opened up new fields of data for mining sentiment, mood, and affective response, leading to the popularity of so-called sentiment analysis. In other words, mood is an important realm of data collection and monitoring for a range of applications and institutions, underwriting the ongoing process of "giving populations over to being a probe or sensor."[30]

The combination of smart phones' monitoring capabilities with the resurgent interest in affective interfaces has led, unsurprisingly, to attempts to use smartphones as mood detectors. Microsoft researchers, for example, have developed a smartphone software system, MoodScope, that infers users' emotional states by monitoring usage patterns. After a two-month period of tracking users and correlating use with mood self-reporting, the researchers claimed to be able to infer the mood of individual users accurately 93 percent of the time.[31] The system is based on the fact that, as the researchers put it, "people use their smartphone differently when they are in different mood states. MoodScope attempts to leverage these patterns by learning about its user and associating usage patterns with particular moods."[32] The virtue of their approach, as the researchers describe it, echoes some of the key attributes of the Cell-All program: it runs in the background and operates passively without users having to input mood information or provide any data other than that generated by the normal daily use of their phones: "[t]he smartphone component runs silently as a background service, consuming minimal power and does not impact other smartphone applications."[33] In other words, its sensing functions become integrated into the fabric of daily life. Locative media like cell phones and tablets are crucial to this endeavor because they "follow" individual users and provide the means for uniquely identifying them, even if they are not named. It would be impossible to use a landline telephone for the type of tracking envisioned by mood probes because the device in question must trace the unique activity patterns of individual users throughout the course of their daily lives.

The stated goal of the development of mood-monitoring capabilities is to assist communication by adding mood awareness to mediated forms of interaction and to assist in data sorting and filtering—that is, helping to provide users with services and communications that are targeted not just to their behavior, to their interests, and to their location but also to their mood. As the Microsoft researchers put it, when mood is integrated into mediated forms of communication, such as e-mail and texting platforms (and even telephony), "[u]sers would be able to better know how and what to communicate with others."[34] They imagine the possibility that mood data will be made available to interlocutors prior to their engaging in communication: we will know whether a friend is sad so that we can call to cheer him up; we will know whether a supervisor is in a bad mood as we craft an e-mail response to him, and so on. In an era of media overload, mood serves as one more input into filtering systems that help us navigate the information landscape. With MoodScope, for example, a "video search could filter results to best match the user's current mood."[35] Of course, marketers will be interested in mood data as well because, as the Microsoft researchers put it, mood "plays a significant role in our lives, influencing our behavior, driving social communication, and shifting our consumer preferences."[36] The fantasy of locative-media-enabled mood tracking, to put it somewhat more bluntly, is to mobilize the "affective register as a more effective avenue of influence.

DRONE THEORY

This section of the chapter draws on the invocation of the figure of the drone as an icon of locative media to consider recent theoretical developments illuminating the forms of knowing associated with data collection in the era of distributed, mobile, interactive devices. Take, for example, the forms of mood monitoring described in the previous section: the incunabulum of the era of so-called affective computing might be described as the attempt to imagine the devices with which we interact as increasingly "human"—not cold calculating machines but communicative prostheses that can respond to a growing range of feelings, moods, and emotions. Once upon a time, developments in machine learning and artificial intelligence were framed in these terms: as the attempt to tear down the distinctions that have been used to separate human from machine—to imagine how computers might take on the attributes of human-ness. Recent developments in theory, however, push in a different, antianthropocentric direction, highlighting the ways in which humans can be understood more broadly as one more collection of things in the great object world. This leveling move is sometimes treated as a form of universal democratization: why should people imagine they are special from the perspective of being (with the associated observation that this sense of privilege might contribute to the mistreatment of other beings in the world, including elements of their physical environment)? Jane Bennett's version of "vibrant" materialism, for example, extends agentic capacity to things with the goal of including "nonhumans in the demos."[37] Ian Bogost's version of "object oriented ontology" performs a similar leveling of being in his embrace of a "flat ontology" that espouses "the abandonment of anthropocentric narrative coherence in favor of worldly detail."[38]

The result might be described as a depsychologizing thrust that works to overcome the opposition between subject and object from the side of the object. Bennett, for example, intentionally gives "short shrift" to subjectivity in order to focus on the "active powers issuing from non-subjects."[39] The antianthropocentric goal is to displace the human subject in favor of "agentive" capacity—something that objects can have as well, according to Bennett, insofar as they can interact with one another in ways that have consequences. Bogost retains the subject but redefines it into oblivion: "The philosophical subject must cease to be limited to humans and things that influence humans. Instead it must become everything, full stop."[40] What falls by the wayside is clear: any psychologized version of subjectivity and thus any meaningful conception of desire, and along with it, any "narrowly human" conception of experience. Bogost's antinarrativizing thrust bears a certain affinity to Bennett's passion for complexification: the familiar accounts of social theory tend to stop too soon, not simply because they focus on human subjects and their interactions but because once you expand the field to include the entire object world, it unfolds indefinitely, vertiginously. No fully accurate account can be provided of anything.

Such approaches, despite whatever progressive claims they might imagine themselves to license, neatly parallel the antipsychologistic, antinarrative logic of drone sensing, and the analytic operations of the data mine. Indeed, Bogost's fascination with the endless expansion of detail, illustrated by his passion for lists, rehearses the omnivorous appetite of the database. I am tempted to label such approaches as a form of drone theory because their versions of agency and experience fit so neatly with those associated with a distributed network of data-sensing probes enabled by the developed of locative media technologies. It would be much harder to effectively outline the attributes of drone experience and drone action without the vocabulary provided by such approaches. From a perspective in which experience and action—and the relationship between the two—is desubjectivized, we can trace some of the salient aspects of "drone thinking," broadly construed:

1. The experience of the drone is the experience of things (in both senses), as summed up in Bogost's broadened definition: "The experience of things can be characterized only by tracing the exhaust of their effects on the surrounding world."[41] That is, things can experience other things only by tracing their "exhaust"—and their own experience is simply whatever reaction they have to this exhaust, a reaction that generates further exhaust. All "internal" aspects of experience associated with accounts of intentionality, motivation, and desire, fall by the wayside, in a formulation that recalls Chris Anderson's paean to the power of big data: "Out with every theory of human behavior, from linguistics to sociology. Forget taxonomy, ontology, and psychology. Who knows why people [and things] do what they do? The point is they do it, and we can track and measure it with unprecedented fidelity."[42] Does the case study of the mood-sensing probe belie such an account, insofar as the probes are attempting to get at some kind of inner experience unique to human subjects? Perhaps it does from the point of view of intersubjective human applications (using mood reading to craft an e-mail to an angry boss, for example), but when it comes to marketing security applications, mood as an internal state is merely a placeholder between two correlations that can eventually, logically, be dispensed with. If a particular pattern of smartphone use correlates with a particular mood and if that mood is, in turn, associated with a heightened probability of response to a certain appeal, it will eventually be possible to jump straight from the pattern of use to the change in response. The distinction between experience and reaction collapses.
2. This collapse highlights the postsubjective character of drone "activity." Sensors incorporate drone logic when they embrace the sensibility of the agent, or actant, whose catalyzing activity is framed as having no recourse to internal, psychological states. Bennett's emphasis on the relationship between the actant and its network is recapitulated by the reminder that the drone/probe must be understood not in isolation but in relation to the arrangements of links, data analytics, and control systems within which it is embedded. As a U.S. Air Force general who oversaw the implementation of the Predator drones put it, "It's about the datalink, stupid."[43] For Bennett, the appeal of the *complexity* of the agent–network relationship is its emergent character and thus the challenge it poses to anthropocentric forms of intentionalism.

> Human actors may well intend a certain outcome, but these intentions become caught up in networks they cannot master, generating unpredictable, unanticipated, and unintentional outcomes.

The goal of the sensor network I have been describing, by contrast, is to harness emergence by capturing its productivity and placing this in the service of defined ends. Mobile phone networks that collect data from increasingly sophisticated sensor arrays will certainly generate more information than can be made sense of by any individual or group, and in this sense they will come to partake of complex arrangements that are, as one recent book put it, "too big to know."[44] The legal theorist Tal Zarsky describes the decisions based on such data-mining processes as "non-interpretable" (and thus nontransparent) because of their inherent complexity: "A non-interpretable process might follow from a data-mining analysis which is not explainable in human language. Here, the software makes its selection decisions based upon multiple variables (even thousands)."[45] However, too big to know does not mean too big to use. Automated data collection and automated sense-making go hand in hand: when too much data to be absorbed or processed by humans can be collected and stored, the promise of data mining is to make the data useful by discerning otherwise inaccessible but nonetheless useful patterns. The cell phone probe becomes one more agent in a burgeoning and distributed network whose emergent properties can be used to guide and monitor centralized forms of response. The elements of what I am here describing as drone theory come to bear on the strategies of information use associated with the huge amounts of data generated by the proliferation of locative media. Put somewhat differently, the invocation of the figure of the drone serves as shorthand for referring to the automated forms of information collection, processing, and response associated with the weaving of mobile devices into the fabric of daily life. If one of the key "advances" of the development of locative media is increased specification and thus enhanced data collection and targeting, such advances simultaneously pose the problem of information glut. More specific information about more people more of the time in a growing range of places means, quite simply, a quantum leap in the amount of data collected. The result, as the figure of the drone implies, is the need for increased forms of automation and thus for a shift to the modes of postnarrative, postinterpretive forms of information use with which these are associated. Both Bennett and Bogost gesture in this direction—albeit unintentionally—insofar as they are fascinated with the proliferation of details and accounts that threaten to overwhelm conventional forms of sense-making. If the thrust of Bennett's account is to highlight how a sensitivity to complex networks of human and nonhuman "agents" tends to swamp any individual intention, the goal of data mining is, in a sense, to domesticate the complexity, subordinating it to determinate goals: policing, security, profit. The drone network, while participating in the logic of emergence, does not succumb entirely: the data mine does not generate its own imperatives. These are imported from elsewhere and imposed by those who control the databases and networks: the datalink, or backend, of the drone array. The era of automated data mining thus augers what might be described as the "drone divide" between those who operate and control the network infrastructure enabling the convenience and ubiquity of locative media and those who carry the sensors with them, generating data as they go.

NOTES

1. Alex Rivera, quoted in Malcolm Harris, "Border Control," *The New Inquiry*, 6 (July, 2012) "Game of Drones," 7.
2. Kashmir Hill, "How to Opt Out of AT&T's Plan to Sell Everything It Knows About You and Your Smartphone Use," *Forbes* (July 3, 2013), accessed August 2, 2013, www.forbes.com/sites/kashmirhill/2013/07/03/how-to-opt-out-of-atts-plan-to-sell-everything-it-knows-about-you-and-your-smartphone-use/.
3. Alexis Kleinman, "Verizon Selling Customers' Cell Phone Data: Report," *The Huffington Post* (May 22, 2013), accessed August 10, 2013, www.huffingtonpost.com/2013/05/22/verizon-selling-customer-data_n_3320680.html. In 2006, a whistle-blower revealed that AT&T was allowing the NSA to tap all communications flowing over its networks, See Ryan Singel, "Spying in the Death Star," *Wired* (May 10, 2007), accessed August 10, 2013, www.wired.com/politics/onlinerights/news/2007/05/kleininterview.
4. Jane Mayer, "What's the Matter with Metadata?" *The New Yorker* (June 6, 2013), accessed September 2, 2013, www.newyorker.com/online/blogs/newsdesk/2013/06/verizon-nsa-metadata-surveillance-problem.html.
5. James Ball, "Verizon Court Order: Telephone Call Metadata and What It Can Show," *The Guardian* (June 7, 2013), accessed September 2, 2013, www.guardian.co.uk/world/2013/jun/06/phone-call-metadata-information-authorities.
6. Mayer, "What's the Matter with Metadata?"
7. Mike Dano, "AT&T Prepping Sale of Customers' Anonymous Location Information and Web, App Usage Data," *FierceWireless* (July 2, 2013), accessed September 2, 2013, www.fiercewireless.com/story/att-prepping-sale-customers-anonymous-location-information-and-web-app-usag/2013–07–02.
8. Kashmir Hill, "Verizon Very Excited That It Can Track Everything Phone Users Do and Sell That to Whoever Is Interested," *Forbes* (October 17, 2012), accessed September 2, 2013, www.forbes.com/sites/kashmirhill/2012/10/17/verizon-very-excited-that-it-can-track-everything-phone-users-do-and-sell-that-to-whoever-is-interested/.
9. Hill, "Verizon Very Excited."
10. Hill, "Verizon Very Excited."
11. Hill, "Verizon Very Excited."
12. "Jargon Watch," *Wired Magazine* 14.02 (February 2006), accessed October 30, 2013, www.wired.com/wired/archive/14.02/start.html?pg=11.
13. Bryan Glyn Williams, "Inside the Murky World of 'Signature Strikes' and the Killing of Americans with Drones," *The Huffington Post* (May 31, 2013), accessed September 2, 2013, www.huffingtonpost.com/brian-glyn-williams/nside-the-murky-world-of-_b_3367780.html.
14. It is worth noting that, at least according to one account, the addition of missile capability to drones was, in effect, an afterthought. The goal was to develop a long-range, always-on surveillance system that allowed for detailed

monitoring of the rhythms of life of potential targets: "Missiles were mounted on Predators only because too much time was lost when a fire mission had to be handed off to more-conventional weapons platforms—a manned aircraft or ground—or ship-based missile launcher. That delay reduced or erased the key advantage now afforded by the drone" (Mark Bowden, "The Killing Machines," *The Atlantic* (August 14, 2013), accessed September 2, 2013, www.theatlantic.com/magazine/archive/2013/09/the-killing-machines-how-to-think-about-drones/309434/). In practice, the attempt to link monitoring with action is one of the elements of what I have called "drone logic"—largely for the practical reasons described in the case of Predator drones: the goal is to make monitoring and response converge in one device or at least in the same network.

15. Kevin Haggerty and Richard V. Ericson, "The Surveillant Assemblage," *The British Journal of Sociology* 51, no. 4 (2000): 605–622.
16. This tendency continues the revolutionary rhetoric of the 1994 "Magna Carta for the Knowledge Age," declaring the "central event of the 20th Century" to be the "overthrow of matter" and signed by noted futurists Esther Dyson, George Gilder, George Keyworth, and Alvin Toffler. Esther Dyson, "Cyberspace and the American Dream: A Magna Carta for the Knowledge Age," *The Information Society* 12, no. 3 (1996): 295–308, www.pff.org/issues-pubs/futureinsights/fi1.2magnacarta.html.
17. Department of Homeland Security, "Cell-All: Super Smartphones Sniff Out Suspicious Substances" (n.d.), accessed September 2, 2013, www.dhs.gov/cell-all-super-smartphones-sniff-out-suspicious-substances.
18. Department of Homeland Security, "Cell-All."
19. Department of Homeland Security, "Cell-All."
20. Department of Homeland Security, "Cell-All."
21. Bill Gates, *The Road Ahead* (New York: Penguin, 1996); Department of Homeland Security, "Cell-All."
22. Department of Homeland Security, "Cell-All."
23. Department of Homeland Security, "Cell-All."
24. Rita Boland, "A Sensor in Every Pocket," *Sensor* 64 (August 2010), 53–54.
25. Rita Boland, "A Sensor in Every Pocket."
26. Neal Lathia, Veljko Pejovic, Kiran K. Rachuri, Cecilia Mascolo, Mirco Musolesi, and Peter J. Rentfrow, "Smartphones for Large-Scale Behaviour Change Interventions," *IEEE Pervasive Computing* (2013), accessed September 2, 2013, http://128.232.0.20/~cm542/papers/ieeepervasive2013.pdf, 1.
27. Evan Perez and Siobhan Gorman, "Phones Leave a Telltale Trail," *The Wall Street Journal* (June 15, 2013), accessed September 2, 2013, http://online.wsj.com/article/SB10001424127887324049504578545352803220058.html.
28. Autotopia Blog, "Cellphone Networks and the Future of Traffic," *Wired* (March 2, 2011), accessed September 2, 2013, www.wired.com/autopia/2011/03/cell-phone-networks-and-the-future-of-traffic/; Jake Sturmer, "Use of Phone-Tracking Technology in Shopping Centres Set to Increase," *ABC News (Australia)* (August 29, 2013), accessed September 2, 2013, www.abc.net.au/news/2013–08–29/use-of-phone-tracking-tech-in-shopping-centres-set-to-increase/4923298.
29. See, for example, Antonio Damasio, *Descartes' Error: Emotion, Reason, and the Human Brain* (New York: Harper Perennial, 1995); Jonah Lehrer, *How We Decide* (New York: Houghton-Mifflin, 2009).
30. Patricia Clough, "The New Empiricism: Affect and Sociological Method," *European Journal of Social Theory* 12, no. 1 (2009): 45.
31. Robert LiKamWa, "MoodScope: Building a Mood Sensor from Smartphone Usage Patterns" (paper presented at the MobySys Conference, June 26, 2013), accessed September 2, 2013, www.ruf.rice.edu/~mobile/publications/likamwa 2013mobisys2.pdf: 1.

32. LiKamWa, "MoodScope," 2.
33. LiKamWa, "MoodScope," 15.
34. LiKamWa, "MoodScope," 3.
35. LiKamWa, "MoodScope," 3.
36. LiKamWa, "MoodScope," 3.
37. Jane Bennett, *Vibrant Matter: A Political Ecology of Things* (Durham, NC: Duke University Press, 2010).
38. Ian Bogost, *Alien Phenomenology: Or What It's Like to Be a Thing* (Minnesota, University of Minnesota Press, 2012).
39. Jane Bennett, *Vibrant Matter*, 44.
40. Bogost, *Alien Phenomenology*, 10.
41. Bogost, *Alien Phenomenology*, 72.
42. Chris Anderson, "The End of Theory: The Data Deluge Makes the Scientific Method Obsolete," *Wired Magazine* (June 23, 2008), accessed August 30, 2013, www.wired.com/science/discoveries/magazine/16–07/pb_theory.
43. Mark Bowden, "The Killing Machines," *The Atlantic* (August 14, 2013), accessed September 2, 2013, www.theatlantic.com/magazine/archive/2013/09/the-killing-machines-how-to-think-about-drones/309434/.
44. David Weinberger, *Too Big to Know* (New York: Basic Books, 2011).
45. Tal Zarsky, "Transparent Predictions," *University of Illinois Law Review* 4 (2013): 1519.

15 Locative Media as Logistical Media

Situating Infrastructure and the Governance of Labor in Supply-Chain Capitalism

Ned Rossiter

INTRODUCTION

The German-based company SAP is one of the largest developers of software that drives global economies, offering leading enterprise software solutions—specifically logistics software—that makes possible movements of people, finance, and things that coalesce as global trade. In its 2012 Annual Report with the not especially modest title, *Helping the World Run Better*, SAP declares that "63% of the world's transaction revenue touches an SAP system."[1] SAP specializes in software development and web-based services associated with Enterprise Resource Planning (ERP) systems in the logistics industries, among many others, including mining, health, finance, medical, insurance, oil and gas, retail, and higher education. This means companies can integrate and automate the majority of their business practices in real-time environments that share common data. So goes the sales pitch.

But how come so little is known about SAP in the fields of media studies, software studies, and network cultures? In part, the answer to this question can be explained with reference to the objects of study that tend to define these fields, especially those of software studies and network cultures that more often valorize open-source initiatives, tactical media interventions, and experimental media culture in general. Less clear is the case of media studies. Broad as the field is, one might expect research into the political economy of media industries to pay some attention to the technology and infrastructure underpinning the global exchange of finance and commodities.[2] Yet this is not the case. Why such an omission looms so large in critical studies of media culture and industries may also have something to do with the fact that logistical software is an aesthetically unattractive and closed proprietary system, even if the logistical infrastructures that are special to transport and communication hold a particular aesthetic allure, which occasionally tips over to the sublime.[3]

The opportunity to begin unraveling the mystery of SAP and its software began to take shape during a recent period of research in Germany during the first half of 2013. Standing around chatting late one evening in the renovated villa accommodating the Medienkulturen der Computersimulation

(MECS) program at Leuphana University, Lüneburg, Christoph Engemann offered an intriguing story to my puzzlement over SAP's supply chain software. He suggested that the origins of SAP's global domination might best be found in the modern rise of double-entry bookkeeping. I sensed a Kittlerian moment in action and found it somewhat unnerving. How was I to possibly trace the media archive of SAP—one that I barely knew beyond a Wikipedia entry—back to a predigital, indeed mediaeval, accounting technology? I was struck by the possibility of an accounting class able to install double-entry bookkeeping as a worldwide institutional standard with the advent of neoliberal capitalism. Former British Prime Minister Margaret Thatcher was presented as the catalyst. But it was the technical capacity of an accounting tool able to filter through corporate and government systems that I found more intriguing.

The rise of the card catalog as a technology of information management could be placed as another reference to the prehistory of SAP software, which drives the world's economies yet remains largely neglected by theorists working in software studies and network cultures.[4] Nowadays, double-entry bookkeeping and file cards seem decidedly quaint when set against the vast array of ERP modules available from logistical software developers such as SAP. Though it is worth noting that SAP does not stand alone here; IBM, Microsoft, and Oracle are other key players in ERP systems, while companies such as Amazon and, up until 2010, Walmart develop their own in-house versions to manage logistical operations. The enormous scope of SAP in terms of organizational culture, product variation, technical systems, economic impact, and the division and multiplication of labor presents numerous challenges in terms of analysis. My interest is to situate SAP within the field of media theory. More ambitiously, I wish to set out a basis for a theory of logistical media.

The larger research, of which this chapter forms a part, will investigate the material dimensions of software systems operative within global logistics industries. That project identifies how software-driven systems generate protocols and standards that shape social, economic, and cross-institutional relations within the global logistics industries. Such operations result in the production of new regimes of knowledge and associated modes of "soft control" within organizational paradigms. The emergent "algorithmic architectures" are computational systems of governance that hold a variable relation between the mathematical execution of code and an "external" environment defined through arrangements of data.[5] The capacity of algorithmic architectures to organize and analyze data on labor productivity in real time, for instance, means that they operate as key technologies for governing labor within logistical industries. My claim is that this has implications for the scope of research on locative media.

This chapter recasts locative media as logistical media. It is interested in how logistical infrastructure is made soft through ERP systems designed to govern the global movement of people, finance, and things. Questions of securitization, control, coordination, algorithmic architectures, protocols, and parameters are among those relevant to a theory of logistical media. The chapter brings logistics, software, and infrastructure together in order to elaborate the conceptual and empirical qualities of what John Durham Peters elusively terms "logistical media."[6] For Peters, the concept of logistical media "stresses the infrastructural role of media."[7] In addition to storage, transmission, and processing systems, I would suggest that the study of logistical media might also include attention to the aesthetic qualities peculiar to the banality of spreadsheets, ERP systems, and the software applications of locational technologies more broadly. The combinatory force of logistical media has a substantive effect on the composition of labor and production of subjectivity. The flexibility of global supply chains and just-in-time modes of production shape who gets employed, where they work, and what sort of work they do. Logistical systems, in other words, govern labor. Logistical labor emerges at the interface between infrastructure, software protocols, and design, and labor situated across global supply chains. The chapter therefore requires an analysis of how labor is organized and governed through software interfaces and media technologies that manage what anthropologist Anna Tsing identifies as "supply chain capitalism."[8]

LOCATING LOGISTICAL MEDIA

Locative media are media of logistics. Logistical media consist of various locational devices such as voice picking technology, GPS tracking, and Radio Frequency Identification (RFID) tags. The spatial and temporal properties of these information and communication technologies have shaping affects on the production of subjectivity. Their primary function is to optimize the productivity of living labor and supply chain operations. Logistical media, as both technologies and software, are very much about locational devices and algorithms that coordinate and control the movement of people, finance, and things. Yet at the conceptual and empirical levels, research on locative media has next to nothing to say about logistical media and supply chain operations whose spatial-temporal operations are frequently overseen by locative media—GPS, RFID, voice picking technology, ERP systems, social media software, and so on.

The deployment of these technologies across logistical supply chains produces what Anja Kanngieser calls "microtechnologies of surveillance" designed "to track and trace workers by constantly tying them to territorial and temporal location[s]."[9] From the embedding of RFID microchips under the skin of employees to the automated instructions on picking lists for workers in warehouses and distribution centers, the use of locational devices within logistical industries results in the extraction and relay of data that holds high commercial value.[10] While geodata may be used in positive ways in the case of managing delivery fleets aimed at fuel efficiency and "ecorouting," locative media also generate data that affects how workers are monitored in workplace settings. Along with privacy issues that arise with the tracking of consignments in transport industries via GPS and cell phones that make visible in real time the location of workers, there is also concern by unions over how the software parameters of voice picking technologies and the generation of data by RFID can also result in the profiling and categorization of workers along lines of race and class that may have deleterious effects on employment conditions and prospects in industries that are frequently characterized by insecure modes of work.

Logistical media are also very different from location-based media characterized by the capacity of users to "control and personalize" the borders between public and private spaces.[11] The agency afforded to users of locative media is much less clear in the case of logistical media, which, as an instrumentalization of location-aware mobile technologies, are designed to exert control over the mobility of labor, data, and commodities as they traverse urban, rural, atmospheric, and oceanic spaces and traffic through the circuits of databases, mobile devices, and algorithmic architectures. A further distinction between locative and logistical media is marked by the tendency of users of locative media to search urban spaces for services related to consumption, while logistical media provide the very conditions for urban settings to function in such a way.

Before moving on to a discussion of logistical media theory as it relates to SAP, some overview of the rise of logistical regimes is required. To date, the study of logistics has largely been undertaken by researchers working in the fields of business and management studies,[12] military history,[13] and economic geography.[14] With military origins, logistics emerged during the Napoleonic wars (1803–1815) as a forecasting technology in the art of warfare, complementing the limits of strategy and tactics.[15] Earlier, in the seventeenth and eighteenth centuries, logistical oversight of supply lines enabled military planners to overcome the practices of pillage and plunder, which kept troops constantly on the move, always in search of food, water, and animal fodder. Logistical operations transformed this nomadic condition, allowing battle to become entrenched around the infrastructure of fortified towns and more sedentary as provisions, troops, and munitions were transported to the front lines of conflict.[16] From its outset, then, "logistical rationality" approached the management of labor through systems of command and control. However, this chapter's point of departure is not focused on the military-industrial complex so much as the interface between infrastructure, software protocols and design, and labor situated across global supply chains (shipping, rail and road transport, procurement, warehousing, IT R&D).

Modern logistics turns around the battle of standards that accompanied containerization in the maritime industries. The standardization of shipping containers from the 1950s was accompanied by disputes among engineers, corporations, and governments over competing economic and geopolitical interests in the transport industries.[17] As geographer Deborah Cowen notes, "Containerization radically reduced the time required to load and unload ships, reducing port labor costs and enabling tremendous savings for manufacturers who could reduce inventories to a bare minimum."[18] By the 1970s, a global standard in containerization had been established, around the same time economic globalization came into full swing following the end of the Bretton Woods Accord in 1971 and the oil crisis of 1973.[19] Sociologists Edna Bonacich and Jake Wilson date what they call the "logistics revolution" from the 1970s, with a particular emphasis on the Reagan and Thatcher eras of market and institutional deregulation, along with neoliberal international free trade agreements.[20] They characterize this organizational revolution in terms of the rise of retailer power over producers and manufacturers in conjunction with changes in production (flexibility and outsourcing), logistics ("intermodalisation" and freight distribution), and labor (intensification of contingency, weakening of unions, racialization of labor, lower labor standards).[21]

Once the logistical problem of container standardization had been resolved, logistical operations shifted attention to the problem of data standards. Although present in a rudimentary form in the 1960s with earlier incarnations in World War II, the computerization of transport industries did not take off until the 1980s, following the standardization of shipping containers and the advent of just-in-time production.[22] Aided by software applications and database technologies, logistics aims to maximize efficiencies at all levels. In other words, the labor control regime is programmed into the logistics chain at the level of code. Similarly, the governance of labor is informatized in such a way that the border between undertaking a task and reporting its completion has become closed or indistinct. As such, there is no longer a temporal delay between the execution of duties and their statistical measure.

In terms of labor management, the optimum state of governance arises at the moment in which the execution of a task, or Standard Operating Procedure (SOP), is registered in the real-time computation of Key Performance Indicators (KPIs). As Katie Hepworth writes, "KPIs and the real-time measurement of labour imply a constant acceleration described in terms of improved productivity."[23] Yet logistics is not bound to the pursuit of speed. The temporal horizon of maritime industries, for example, may just as often require a slowing down of movement or even periods of stasis.[24] The capacity to calibrate time according to multiple and frequently conflicting economic interests constitutes a form of "transactional impedance [that] relates to the power-geometries within supply chains."[25] As Hepworth explains, "It describes how diverse stakeholders and individuals are placed in quite distinct relations to commodity flows and the interconnections between global sites, as well as the ways in which these relations are manipulated for the benefit of particular actors within those networks."[26] Logistical software, ERP systems, and technologies of location play a central role in the mediation of such economies of transaction and data extraction.

"WE HELP RUN THE WORLD BETTER AND IMPROVE PEOPLE'S LIVES" (SAP)

SAP's ERP and logistics software generally remains a black box to most. Even those who use the software have little idea of how it works. For this reason, SAP's supply chain software can be considered a form of imaginary media.[27] You choose what you want to put into that space because if you don't, then others certainly will. On this note, we might ask how much—or to what extent—does imagination shape protocols? One recent tech report went so far as to say SAP "poisons networks." And then continued with this little gem: "SAP's expensive business software, which no one knows what it does, and is so esoteric that no one ever bothers to upgrade it, could be a ticking security bomb."[28] So what is supply chain software, and what does it do? Why don't we have it installed on our PCs and laptops? Why are we so utterly unaware of it?

The digital humanities and software studies may have something to contribute by way of response to these sorts of questions. But both would need to radically shift their focus away from a general mission to digitize the humanities archive and conduct exotic sorties into the fringes of network cultures. These are important enough activities, but they tell us little about how big power works. First of all, we need to enter the imaginary world of SAP. We need to pose critical questions based not on our disciplinary predilections and intellectual whimsies but rather on the object of inquiry—computational power, interface aesthetics (what many these days call "usability," which is so dreary), algorithmic architectures, and the politics of parameters. What is required is a truly transdisciplinary collective investigation into the increasingly mysterious centers of power in the age of big data. This would involve work between media theorists, organizational studies, computer scientists, programmers, and designers to open up the black box of SAP and the products of similar software developers, identifying how their algorithmic architectures are constructed, what their business models are, and how they use data extracted from the back end of mostly unwitting clients. What is the vision of SAP beyond the PR machine? According to one SAP consultant, it is 1 billion SAP users by the year 2020. What does Hasso Plattner & Co. see as the limit horizon for extracting value from the world? We need to know that because, whether we are aware of it or not, our lives are becoming increasingly subsumed by logistical nightmares.

SAP is renowned for its real-time ERP software, beginning with the R/1 financial accounting system, which was developed in the early 1970s. This was soon followed in the late seventies by the text-based R/2 software—"a mainframe based business application suite" able to handle multicurrencies and multilanguage requirements. Making use of relational databases such as Oracle (SAP's main competitor), R/3 was launched in 1992 with a graphical user interface based on three-tier client–server architecture able to scale and integrate multiple operating systems.[29] In November 2010, the SAP HANA product was released. This was SAP's response to the challenge of big data. As the product spiel on SAP's site reads: "SAP HANA can help you dramatically accelerate analytics, business processes, predictive analysis, and sentiment data processing—all on a single in-memory computing platform."[30] More recently, SAP is dealing with the computational (security) and market challenge of cloud computing, with ventures into the world of educational courseware such as MOOCs (massive open online courses).

One of the keys to SAP's success has been the development of modules within the ERP systems that can handle multiple aspects of business operations. Some of these modules include human resource management, finance and accounting, sales and distribution, production planning, warehousing, procurement, supply chain management, and logistics. SAP's money is made through a combination of license fees, consultation, customized implementation of ERP packages, and ongoing maintenance of modules within the operational context of its clients. The cost for companies is enormous, reaching into the hundreds of millions for large corporations and into the tens of millions for small and medium-sized enterprises (SMEs). The exact cost for companies varies considerably, with pricing made highly flexible and negotiable depending on the extent of customization required and the geographic and national location of the organization wishing to adopt SAP's ERP system. Ten years ago, SAP education and training was another profitable line of revenue, but since the global financial crisis, hotels have been empty for SAP education sessions.[31]

Through its widespread application in a variety of industries with around 29 million users worldwide, SAP software not only shapes business practice but profoundly affects people and the planet. Along with providing the dominant interface for managing our global economies engaged in planetary obliteration, SAP scrutinizes the experience and conditions of labor in the fairly unforgiving regime of real-time performance measures. It would seem obvious, then, for SAP software to constitute an important object of inquiry not only for business and management or computing but equally for social, political, and cultural analysis. My interest in this chapter, however, is a bit more specific: what does SAP have to contribute to a theory of logistical media?

A theory of logistical media might begin with a critical analysis of software, focusing specifically on how SAP logistics software is designed in ways that govern both labor and global supply chains. Beyond the need for some quite sophisticated programming knowledge, a key obstacle to such an undertaking concerns the prohibitive pricing of SAP software (which runs into the millions to install and maintain), its corporate secrecy, and the commercial value of the data it collects. Despite these very real constraints for software critique, one notable intervention in SAP's prison house of code occurred in 1996 in Dortmund, Germany, when anarchist programmers affiliated with LabourNet—a network of labor unions—cracked SAP software.[32] The motivation, according to LabourNet's Helmut Weiss, was to identify specific lines of code within SAP software that would affect workplace activity in detrimental ways.[33] Algorithmic parameters became the basis for union negotiations.

Yet despite this quite exceptional feat of code breaking, Weiss notes that "[n]ot a single union accepted our proposal (as the media-workers union)" to develop this further as a topic of political research.[34] Weiss went on to explain two key reasons for this disinterest on the part of unions in addressing the technical parameters of labor control that impact on their constituencies: "First the traditional approach of German unionism towards technology—always in favor of new technologies, despite not knowing anything about them. (A position widely shared even by the left wing unionists)."[35] Second: "[t]hey all (at that time it was, practically speaking, mainly the metalworkers union IG Metall and public services union ÖTV who had to deal with SAP) were afraid of 'illegal action' so they distanced themselves from our initiative."[36] Even though the media-workers union had a collective agreement in 1990 on "life long learning," where one of the goals was "to enable a critical analysis of all tools we use while working,"[37] the everyday practice, according to Weiss, never matched up with such a principle, especially as employer associations went on the attack in the early nineties. Over two decades have passed since this earlier attempt to bring a political knowledge of code to address labor conditions for those working in supply chain capitalism. A media theory of logistics, including an account of how infrastructure is made soft, provides one index for reconstituting a political knowledge of what might be termed "logistics in the age of algorithmic capitalism."

SOFT INFRASTRUCTURE

The primary task of the global logistics industry is to manage the movement of people and things in the interests of global trade and the optimization of efficiency. Logistics infrastructure manifests as roads, railways, shipping ports, intermodal terminals, airports, and communications facilities and technologies. Logistics infrastructure enables the movement of labor, commodities, and data across global supply chains. Increasingly, logistics infrastructure is managed through computational systems of code and software. As such, algorithms play a vital role in arranging the material properties and organizational capacities of infrastructure. Algorithms thus register a form of infrastructural power.

One consequence of such computational power can be seen in the way logistical infrastructure not only manifests in particular locations and material forms but has the capacity to scale across geographic registers, technological systems, and social, political, economic, and cultural settings. Scale is calibrated according to real-time systems of measure and performance special to just-in-time regimes of labor productivity and commodity assemblages. Software management systems that oversee transport and communications infrastructure may be located in remote settings at considerable distances from the metropolitan scenes of activity, registering a shift from Saskia Sassen's concept of the "global city" of finance centers to one predicated on the peripheral operations of the "logistical city."[38] It is in this sense that infrastructure takes on a quality of "liquid modernity," making possible organization at a distance.[39] Algorithmic architectures are thus central to logistical operations and indeed to global service-based markets and value-adding systems.

The software applications special to logistics visualize and organize these mobilities, producing knowledge about a world undergoing massive transition into a digital era. Software-driven systems generate protocols and standards that shape social, economic, and cross-institutional relations within the global logistics industries. Such operations result in the production of new regimes of knowledge within an organizational paradigm. The logistical industries that drive supply chain capitalism consist of different infrastructural aspects of the expanding sector of logistics: transport, communications, warehousing, procurement, and ports. Such infrastructure is frequently coded and managed by computational systems. Rob Kitchin and Martin Dodge analyze contemporary urban settings as forms of "code/space" in which "software and the spatiality of everyday life become mutually constituted."[40] The digital coding of space—or the making soft of infrastructure—has impacts on how labor is managed and how subjectivities are produced when the time of life and action of bodies is increasingly overseen and regulated by computational systems of control and regulation.

Soft infrastructure may be understood in four key ways: First, as algorithmic architectures (software platforms and code) operating transport and communications infrastructure that connect global supply chains to the managerial science of logistics. Second, as communications and transport infrastructure that materializes algorithmic agency (or the shaping power of code and software) and serves as a central "problem" for R&D and policy making in the logistics industries. Third, in terms of formal and informal labor sectors that adopt different and often conflicting algorithmic architectures operating within supply chain capitalism. And fourth, as models of governance or "protocological control"[41] that draw on big data to organize and manage the mobility of people, finance, and things within infrastructural systems on local and transnational scales.

Once combined, these features of soft infrastructure provide the basis for developing a theory of logistical media. A theory of logistical media derived from a study of SAP might be organized around the following headings: securitization, coordination, control, simulation, models, algorithmic architectures, the Internet of Things, protocols, standards, and parameters. Although there is no question that logistical systems can facilitate greater workplace productivity and improve supply chain management, it is also the case that the seamless interoperability sought by policy makers and technologists is often disrupted by labor struggles, software glitches, bottlenecks in transport and computer network traffic, infrastructural breakdowns, inventory blowouts, sabotage of supply chains, and the like.

LOGISTICAL MEDIA THEORY

The history of materialist approaches to the study of communication is one obvious point of departure for a media theory of logistical systems of communication, coordination, and control.[42] A focus on the material properties special to transport and communications infrastructure can be analytically complemented with attention to how the algorithmic architectures of communication and transport infrastructure impact on the experience and conditions of labor operating within those industries. Drawing on the work of medium theorists such as Innis, McLuhan, and Ong, communication historian and cultural critic James Carey noted that the advent of telegraphy in the nineteenth century "freed communication from the constraints of geography."[43] This meant that the concepts and practices of communication could be understood beyond the dominant "transmission view" of communication in which the mobility of people, goods, and information involved equivalent operations (using railway networks, for example). For Carey, symbolic and ritualistic views of communication were able to develop.

But as a number of media and communication scholars working in the tradition of Carey and medium theorists have recently observed, the history of mobile technologies demonstrates an ongoing linkage with transportation technologies.[44] As Mimi Sheller explains, "the advent of mobile communication technologies and software-supported transportation networks also fundamentally changes how communication is thought about, but in this case by re-embedding it into transportation infrastructures and spaces of transit, which are also spaces of transmission."[45] Lisa Parks notes that the term "infrastructure" emphasizes the materiality and distribution of communication.[46] It also reminds us of the territoriality and geography of communication and transport, of questions of power, and of the challenge to devise new techniques and modes of visualizing these interrelations.[47]

Software studies, as it has emerged from the study of network cultures and critical studies of digital media, is another key field for developing a theory of logistical media. Although much more attuned to the work of critique and an often high technical knowledge of digital media, software studies tend to focus on open source software and investigate questions of materiality in terms of design and "cultural analytics," "protocological control," "media ecologies," "memory and storage," and "media archaeology."[48] Combining empirical study with concept development constitutes an intervention within the emergent field of software studies by shifting the analytical gaze from open source software cultures and "cultural analytics" to where the vapor-ware meets the hard edge of consultancy culture and global infrastructures.[49] In doing so, the closed, proprietary systems of software that manage global supply chains have a substantive impact on modes and practices of work.

The scalar dimension of software, for example, is dependent on the interoperability of protocols and the hegemony of standards. As David Dixon, the SAP program manager for the UK supermarket chain Asda (a subsidiary of Wal-Mart), notes, "The main rationale [behind the SAP rollout] was to have a 'one version of the truth' approach, to standardize and gain a degree of control around the world."[50] Put another way, the market penetration of software is without doubt shaped by its capacity to communicate with a wide range of software applications and hardware devices. Of course, this is only part of the story; nevertheless, interoperability is key to the political economy of software. Once universalized across the vertical distribution of organizations, from large corporations to SMEs, an ERP system designed by a developer such as SAP has the power to determine whom you do business with by the fact that transactions are simply easier when your trading partner is on the same platform. In other words, a monopoly effect arises from the trans-scalar integration of ERP systems across organizational settings. This may seem to be overstating the fact of interoperability among competing ERP systems, but from what I have been told by various people working in the world of SAP, the tendency is for companies to give preference to other businesses also on the SAP system. Indeed, companies are encouraged by SAP to spread the good word of SAP. I have no idea what sort of commission or negotiated adoption fees companies might attract for such advocacy work.

Enabling the communication of objects, the Internet of Things (IoT) has also become central media components to the logistical industries.[51] Consisting of RFID tags, sensors, 3D printing, mobile devices, and software or robotic actuators, the IoT and expansion of communication standards offers a network effect to the silo models of machine-to-machine communication (M2M). New regimes of value and the scalability of data are key attractions that the IoT brings to logistical operations, although for a company like SAP the communication of objects would be likely to occur not over the public Internet but rather through private networks in the interests of data securitization and economies of scarcity. For advocates of a public IoT, this makes the battle over open source standards a central issue. Without them, the capacity for trans-scalar and multiplatform interoperability would be severely circumscribed. As Fenwick McKelvey, Matthew Tiessen, and Luke Simcoe note:

> The growing mediation of everyday life by the Internet and social media, coupled with Big Data mining and predictive analytics, is turning the Internet into a simulation machine. The collective activity of humanity provides the data that informs the decision making processes of algorithmic systems such as high-frequency trading and aggregated

news services that, in turn, are owned by those who wield global power and control: banks, corporations, governments.[52]

At stake here is the question of accessibility to communications infrastructure, or what Sheller terms "new forms of infrastructural exclusivity,"[53] once logistical firms such as SAP become the drivers of developing the Internet of Things. We might also ask what sort of simulation machine do we wish to inhabit?

CONCLUSION

In further developing a logistical media theory, I would suggest that three key dimensions of the materiality of communication might serve as framing devices for ongoing research. First, *the materiality of concrete things* (the infrastructure of ports, IT zones, rail and road transportation, container yards, warehouses). Second, *the materiality of communication itself* (the spatial, temporal and aesthetic properties of digital communication technologies and software). And third, *the materiality of practices that condition the possibility of communication* (the labor of coding and design in developing algorithmic architectures, coupled with labor experiences and conditions across sectors within the logistics industries).[54]

Logistical media theory, understood within these analytical and material coordinates, also holds relevance for research into both the conceptualization and analysis of practices special to locative media. Whether it is RFID tags, GPS devices, or voice picking technologies, these ubiquitous media of location associated with the Internet of Things assume—like logistics—a world of seamless interoperability. But we know this to be a fantasy of technologists, policy makers, and advertising agencies. Struggles over communication protocols, infrastructural standards, mobile populations, and expressions of refusal by labor comprise just some of the glitches that always accompany the operation of locative media as logistical media.

NOTES

1. SAP, *Helping the World Run Better*, 2012 Annual Report: 4, accessed August 27, 2013, www.sapintegratedreport.com/2012/fileadmin/user_upload/2012_SAPintegratedreport/downloadcenter/SAP_AR2012_en.pdf.
2. An exception to these tendencies can be found in the adjacent field of science and technology studies (STS). See Neil Pollock and Robin Williams, *Software and Organisations: The Biography of the Enterprise-Wide System or How SAP Conquered the World* (Oxon and New York: Routledge, 2009). This book only came to my attention after writing this chapter.
3. See the photographs and documentary films of Edward Burtynsky, for example: *Manufactured Landscapes* (Ottawa and New Haven: National Gallery of Canada and Yale University Press, 2003); *China: The Photographs of Edward*

Burtynsky (London: Steidl, 2005). See also Soenke Zehle, “Dispatches from the Depletion Zone: Edward Burtynsky and the Documentary Sublime,” *Media International Australia* 127 (May 2008): 109–115.

4. See Markus Krajewski, *Paper Machines: About Cards & Catalogues, 1548–1929*, trans. Peter Krapp (Cambridge, MA: MIT Press, 2011).
5. See Luciana Parisi, “Algorithmic Architecture,” in *Depletion Design: A Glossary of Network Ecologies*, eds. Carolin Wiedemann and Soenke Zehle (Amsterdam: XMLab and the Institute for Network Cultures, 2012), 7–10. See also Parisi’s *Contagious Architecture: Computation, Aesthetics and Space* (Cambridge, MA: MIT Press, 2013).
6. John Durham Peters with Jeremy Packer, “Becoming Mollusk: A Conversation with John Durham Peters About Media, Materiality and Matters of History,” in *Communication Matters: Materialist Approaches to Media, Mobility and Networks*, eds. Jeremy Packer and Stephen B. Crofts Wiley (New York: Routledge, 2012), 35–50.
7. Peters and Packer, “Becoming Mollusk”: 43.
8. Anna Tsing, “Supply Chains and the Human Condition,” *Rethinking Marxism* 2 (2009): 148–176.
9. Anja Kanngieser, “Tracking and Tracing: Geographies of Logistical Governance and Labouring Bodies,” *Environment and Planning D: Society and Space* 34, no. 4 (2013): 598. The remainder of my account of RFID, GPS, and voice-picking technologies in this paragraph draws on research in Kanngieser’s article.
10. The question of authorship and ownership of data produced through the use of locative media is one raised in an article written around the period that saw the first wave of critical research on locational-aware technologies. See Anne Galloway and Matt Ward, “Locative Media as Socialising and Spatializing Practice: Learning from Archaeology,” *Leonardo Electronic Almanac* 14, nos. 3–4 (2006), accessed August 27, 2013, www.leoalmanac.org/leonardo-electronic-almanac-volume-14-no-3-4-june-july-2006/.
11. Adriana de Souza e Silva and Jordan Frith, “Location-Aware Technologies: Control and Privacy in Hybrid Spaces,” in Packer and Wiley, *Communication Matters*, 266.
12. See Martin Christopher, *Logistics and Supply Chain Management*, 4th ed. (Harlow, UK: Pearson, 2011); Donald Waters, ed., *Global Logistics: New Directions in Supply Chain Management* (London: Kogan Page, 2010).
13. Martin van Creveld, *Supplying War: Logistics from Wallenstein to Patton* (Cambridge: Cambridge University Press, 1977); Paul Virilio, *War and Cinema: The Logistics of Perception* (London: Verso, 1989).
14. Deborah Cowen, “A Geography of Logistics: Market Authority and the Security of Supply Chains,” *Annals of the Association of American Geographers* 100 (2010): 600–620; Craig Martin, “Desperate Mobilities: Logistics, Security and the Extra-Logistical Knowledge of ‘Appropriation.’” *Geopolitics* 17, no. 2 (2012): 355–376.
15. See Carl von Clausewitz, *On War* (Oxford: Oxford University Press, 2007). See also Brett Neilson, “Five Theses on Understanding Logistics as Power,” *Distinktion: Scandinavian Journal of Social Theory* 13, no. 3 (2012): 323–340.
16. See Manuel DeLanda, *War in the Age of Intelligent Machines* (New York: Zone Books, 1991).
17. See Marc Levinson, *The Box: How the Shipping Container Made the World Smaller and the World Economy Bigger* (Princeton, NJ: Princeton University Press, 2006).
18. Cowen, “A Geography of Logistics,” 612.

19. See David Harvey, *The New Imperialism* (Oxford: Oxford University Press, 2003).
20. Edna Bonacich and Jake B. Wilson, *Getting the Goods: Ports, Labor and the Logistics Revolution* (Ithaca, NY: Cornell University Press, 2008).
21. Bonacich and Wilson, *Getting the Goods*, 6–22.
22. See Brian Holmes, "Do Containers Dream of Electric People: The Social Form of Just-in-Time Production," *Open* 21 (2011): 30–44. See also Brian Ashton, "Logistics and the Factory Without Walls," *Mute Magazine* (September 14, 2006), accessed August 27, 2013, www.metamute.org/editorial/articles/logistics-and-factory-without-walls.
23. Katie Hepworth, "Interventions Towards the Logistical City," *Environment and Planning D: Society and Space* (forthcoming, 2014).
24. See Brett Neilson and Ned Rossiter, "Waiting, Still Moving: On Migration, Logistics and Maritime Industries," in *Stillness in a Mobile World*, eds. David Bissell and Gillian Fuller (London and New York: Routledge, 2011), 51–68.
25. Hepworth, "Interventions Towards the Logistical City."
26. Hepworth, "Interventions Towards the Logistical City."
27. See Erik Kluitenberg, ed., *Book of Imaginary Media: Excavating the Dream of the Ultimate Communication Medium* (Rotterdam: NAi, 2006).
28. Nick Farrell, "Ancient SAP Software Poisons Networks," *TechEye* (June 18, 2013), accessed August 19, 2013, http://news.techeye.net/security/ancient-sap-software-poisons-networks.
29. Technical details taken from the "SAP R/3" entry on Wikipedia, with additional reference to the SAP Community Network thread on R/2 and R/3: Venkatesh Rajendran, "Can You Tell Me Some Differences Between R2 and R3 Systems?" *SAP Community Network* (December 20, 2006), accessed August 19, 2013, http://scn.sap.com/thread/282002.
30. "Invent New Possibilities with SAP HANA Solutions," In-Memory Computing (SAP HANA) (2013), accessed August 19, 2013, www.sap.com/solutions/technology/in-memory-computing-platform/hana/overview/index.epx.
31. Details on these SAP operations have been gleaned from discussions in Germany in 2013 with professionals working in SAP securitization, market analysis, and global project management. In order to obtain some insight into the corporate practices of SAP and its software interface for supply chain coordination, organizational culture, data economies, and event processing, further discussions about the SAP Human Capital Management (HCM) Module where held with SAP consultant Michael Hellmich, followed by an introduction to the SAP Supply Chain Management (SCM) Modules by Anselm Roth, a solution architect at SAP Germany. These formed part of an international workshop on SAP, software, and labor that I organized with Götz Bachmann, Armin Beverungen, and Timon Beyes at Leuphana University's Centre for Digital Cultures: "Logistics of Soft Control: SAP, Labour, Organization," Lüneburg (June 20–21, 2013), https://www.leuphana.de/zentren/cdc/aktuell/termine/archiv/ansicht/datum/2013/06/20/workshop-logistics-of-soft-control-sap-labour-organization.html.
32. LabourNet Germany (2013), accessed August 19, 2013, www.labournet.de/.
33. Many thanks to Helmut Weiss for his discussions during the Leuphana workshop on SAP and for this account, in particular, of union research in IT within Germany.
34. Helmut Weiss, e-mail communication, July 9, 2013.
35. Helmut Weiss, e-mail communication, July 9, 2013.
36. Helmut Weiss, e-mail communication, July 9, 2013.
37. Helmut Weiss, e-mail communication, July 9, 2013.

38. Saskia Sassen, *The Global City* (Princeton, NJ: Princeton University Press, 1991). For a discussion of the concept and condition of the logistical city, see Ned Rossiter, "Logistical Worlds," *Cultural Studies Review* (forthcoming, 2014). See also Brett Neilson and Ned Rossiter, "The Logistical City," *Transit Labour: Circuits, Regions, Borders* 3 (August 2011): 2–5, http://transitlabour.asia/documentation/.
39. Zygmunt Bauman, *Liquid Modernity* (Cambridge: Polity, 2000).
40. Rob Kitchin and Martin Dodge, *Code/Space: Software and Everyday Life* (Cambridge, MA: MIT Press, 2012).
41. Alexander R. Galloway, *Protocol: How Control Exists After Decentralization* (Cambridge, MA: MIT Press, 2004).
42. See Harold A. Innis, *The Bias of Communication* (Toronto: University of Toronto Press, 1951); Lewis Mumford, *Technics and Civilization* (New York: Harcourt, Brace & Co., 1934); Manuel Castells, *The Rise of the Network Society* (Oxford: Blackwell, 1996); Manuel Castells, *Communication Power* (Oxford: Oxford University Press, 2009); Friedrich A. Kittler, "There Is No Software," in *Literature, Media, Information Systems* (Amsterdam: G+B Arts International, 1997), 147–155; Friedrich A. Kittler, *Gramophone, Film, Typewriter*, trans. G. Winthrop Young and M. Wutz (Stanford, CA: Stanford University Press, 1999); Lisa Parks, *Cultures in Orbit: Satellites and the Televisual* (Durham, NC: Duke University Press, 2005); John Durham Peters, "Calendar, Clock, Tower," in *Deus in Machina: Religion and Technology in Historical Perspective*, ed. Jeremy Stolow (New York: Fordham University Press, 2012), 25–42.
43. James Carey, "Technology and Ideology: The Case of the Telegraph," in *Communication as Culture: Essays on Media and Society* (Boston: Unwin and Hyman, 1989), 204. Cited in Mimi Sheller, "Materializing US–Caribbean Borders: Airports as Technologies of Communication, Coordination and Control," in Packer and Wiley, *Communication Matters*, 233.
44. See Packer and Wiley, *Communication Matters*; David Morley, "For a Materialist, Non-Media-Centric Media Studies," *Television & New Media* 10, no. 1 (2009): 114–116; Gerard Goggin, *Global Mobile Media* (Oxon, UK: Routledge, 2011); Larissa Hjorth, *Mobile Media in the Asia-Pacific: Gender and the Art of Being Mobile* (Oxon, UK: Routledge, 2009); Mark Andrejevic, "Media and Mobility," in *Media History and the Foundations of Media Studies*, *The International Encyclopedia of Media Studies*, Vol. 1, ed. John Nerone (Cambridge and Malden, MA: Wiley-Blackwell, 2013), 521–535.
45. Sheller, "Materializing US–Caribbean Borders": 233.
46. Lisa Parks, "Infrastructural Changeovers: The US Digital TV Transition and Media Futures," in *Media Studies Futures*, *The International Encyclopedia of Media Studies*, Vol. 5, ed. Kelly Gates (Cambridge and Malden, MA: Wiley-Blackwell, 2013), 296–317.
47. Parks, "Infrastructural Changeovers."
48. See, respectively, Lev Manovich, *Software Takes Command* (New York: Bloomsbury Academic, 2013); Alexander R. Galloway, *Protocol: How Control Exists After Decentralization* (Cambridge, MA: MIT Press, 2004); Matthew Fuller, *Media Ecologies: Materialist Energies in Art and Technoculture* (Cambridge, MA: MIT Press, 2005); Wendy Hui Kyong Chun, *Programmed Visions: Software and Memory* (Cambridge, MA: MIT Press, 2011); Erkki Huhtamo and Jussi Parikka, eds., *Media Archaeology: Approaches, Applications and Implications* (Berkeley: University of California Press, 2011).
49. See Lev Manovich, "Cultural Analytics: Visualizing Cultural Patterns in the Era of 'More Media'" (2011), http://manovich.net; Keller Easterling, *Enduring*

Innocence: Global Architecture and Its Political Masquerades (Cambridge, MA: MIT Press, 2005).

50. Quoted in Anh Nguyen, "Walmart Pushes Ahead with SAP Rollout After Asda Pilot Success," *Computer World UK: The Voice of IT Management* (August 23, 2010), accessed August 27, 2013, www.computerworlduk.com/news/it-business/3236348/walmart-pushes-ahead-with-sap-rollout-after-asda-pilot-success/.
51. For an overview of IoT, see Rob van Kranenburg, *The Internet of Things: A Critique of Ambient Technology and the All-Seeing Network of RFID*, Network Notebooks no. 2 (Amsterdam: Institute of Network Cultures, 2008), accessed August 19, 2013, www.networkcultures.org/_uploads/notebook2_theinternetofthings.pdf.
52. Fenwick McKelvey, Matthew Tiessen, and Luke Simcoe, "We Are What We Tweet: The Problem with a Big Data World When Everything You Say Is Data," *Culture Digitally: Examining Contemporary Cultural Production* (June 3, 2013), accessed August 19, 2013, http://culturedigitally.org/2013/06/we-are-what-we-tweet-the-problem-with-a-big-data-world-when-everything-you-say-is-data-mined/.
53. Sheller, "Materializing US–Caribbean Borders": 238.
54. The first two of these categories are adapted from Packer and Wiley, "Introduction: The Materiality of Communication," in *Communication Matters*: 3. The third dimension to the materiality of communication is one that I see as both a precondition for and coincident with the other two.

16 Locative Esthetics and the Actor–Network

Michael Dieter

We frequently encounter locative media today not as experimental artworks or speculative prototypes but as commercial services or obfuscated security systems. From the Apple iPhone to Google Maps, location-based technologies are now deeply integrated into the corporate collection of spatial analytics, along with the widespread privatized annotation of landscapes and urban environments.[1] Parallel to these developments has been what has been described as an increasing militarization of space through location-aware profiling systems and antiterrorist premediation.[2] These transformations in the political economy of mapping and the asymmetries of "dataveillance," taken together, arguably complicate a number of essential ambitions that defined early locative media art, despite this movement also having always held close links with industry and having utilized military-supported devices. To be sure, disentangling such aspirations from the aspects of surveillance, capture, and spectacle was always a major challenge, especially for creative engagements with sociotechnological assemblages that are bound by uneven terrains of power and capital. It has also included recognizing the agential destabilization that follows trends toward ubiquitous computing or the Internet of Things, not only through gadgets but also the embedding of digital sensors, automated computational mapping systems, and algorithmic devices throughout everyday life. These environments suggest a rearrangement of sense, perception, and memory in the spatial deployment of calculative procedures.[3] As N. Katherine Hayles notes, the infrastructural shift raises important ontopolitical concerns because "the relations between human, animal and thing come up for grabs, functioning as a chaotic nexus in which technological innovations, anxieties about surveillance and privacy, capitalistic and military exploitations, and creative storytelling swirl together in a highly unstable and rapidly changing dynamic."[4] One task, according to Hayles, involves somehow conceptually moving beyond discourses based on anthropocentric politics of resistance to rethinking the limits of human subjectivity and behavior for purposes of collective empowerment. This is a critical project because the frequently masked operative scope of these systems from user populations means that changes are often affectively experienced only through a creeping sense of paranoia, rather than being approached directly in conscious and active experimentations. As Eric Gordon and Adriana de Souza e Silva put it, "while we are location aware, we are riddled with anxieties about it."[5]

Just over a decade ago, artistic experimentation with geographic information systems (GISs) precisely expressed desires to restore authentic social relations against the disorientating informational flows of global network societies and to contest concentrations of power through innovative participatory events. The term "locative media" originates from that period, first being coined by Karlis Kalnins at a creative workshop at the Center for New Media in Riga, Latvia.[6] It was originally led by an ambition to "to take art out of the galleries and off the screen"[7] and to reinvigorate avant-gardist aspirations by initiating a "paradigmatic shift" in the spatial esthetics of the everyday.[8] A classic desire for (re)integrating art and life, for instance, could be detected in ongoing references to Situationist techniques of *détournement* and *dérive* as the basis for new experiential modes of inquiry.[9] Researchers like Anne Galloway and Matt Ward, for instance, proposed that archaeological methods, photography, global positioning systems (GPSs), and software-based cartography could be merged as locative media systems, where multidisciplinary projects would reanimate static architecture to restore "hope" in a collective transformation of space.[10] In a similar way, Jeremy Hight outlined how forgotten historical knowledge might be triggered or recovered from the landscape through a new site-specific attentiveness:

> The authoritative voice of intellectual discourse is counter-intuitive to creating works that speak of place, events, moments, important layers of a place lost in time. Instead of the voice of authority, what is needed is the voice of the work and location itself: the information, the artistic use of language and image and most of all, of the agitation into being of a location as multi-tiered, alive.[11]

Forms of transitory civil unrest, protest, and dissent were imagined as being recorded through situated archiving and stored for future access as evidence of political struggle. Such rhetoric conveyed a longing for authentic modes of community that would be guaranteed by a vital combination of new media devices and active participants. Even a libertarian program such as Ben Russell's *Headmap Manifesto*—where wireless technologies were read as fragmenting the nation-state through mutable cartographies—hinged on a close association between "hybrid space" and the possibilities for collective sociopolitical emancipation.[12] In the extreme, independent researcher and curator Bonnie DeVarco argued that the collaborative possibilities of locative media would instantiate a new sense of ecological stewardship by actualizing Buckminster Fuller's utopian World Game project.[13]

It would be unfair to dismiss such imaginings in retrospect as simply naïve; however, it is nevertheless difficult to reconcile such early paradigms with, for instance, the hyperindividualistic user experience of Google, ludic capitalism of platforms like Foursquare, or clandestine schemes of the U.S. National Security Agency (NSA). These developments speak to a major struggle with digital and networked systems: they hold the potential for establishing associated milieus and collective enunciations but frequently resolve into hierarchical and reiterative formations of power.[14] On the one hand, expropriations of data occur through the aggregate, yet processes of valorisation, profiling, and predictive mapping are generated by these technologies precisely in order to feed back along micropolitical dimensions.[15] A variety of possible compositions exist for these infrastructures; however, the potential exacerbation of unequal dynamics must factor into any analysis of the informational production of space today.

Keeping experimental desires in mind for locative arts, this chapter discusses creative mapping projects at the intersection of technical affordances, corporate control, state surveillance, and artistic practice. Here, it should be noted that the liberalization of GPS, the rapid adoption of "smart" location-aware devices, radio frequency identification (RFID), wireless networks, and the general integration of locative data into digital content (by logging IP addresses or indexing geotagged information) has led to several productive reconsiderations of cartographic practice and theory.[16] These emerging frameworks, interestingly, have followed a trajectory from modes of representational analysis and ideology critique, toward understandings of distributed agencies, radical empiricism, and ontogenetic transductions, only to encounter difficulties concerning new entanglements of power. Even though locative media systems are increasingly available to users and programmers, data capture has become an essential characteristic of a sizable sector of closed infrastructural settings and contexts (military, governmental, scientific, corporate).[17] A key quandary is how software models for capturing geospatial data are imbricated with ongoing political conflicts, tensions, inequalities, and other asymmetries, especially when operating as obfuscated commercial and governmental systems built for population-scale explication.

What follows, therefore, is ultimately a consideration of how locative media art might be theorized today as a form of critical technical practice.[18] That is, the approach taken here is a deliberation on how certain artworks are invested in modes of problem creation and discovery within the complexities of net localities and common atmospheres, an aspect that is imperative given the tendency for early discourses on locative media to dwell on pragmatic solutions and technical remedies.[19] My argument, moreover, is an engagement of sorts with theoretical frameworks that borrow from Bruno Latour's actor–network theory (ANT) for practical, conceptual, and interpretative purposes. Indeed, I make the claim that ANT-style accounts could benefit from a more wide-ranging interpretation of esthetic comportments, especially in relation to the contingencies of inequality in the habitual reproduction of locative media. This chapter, therefore, provides an extended reading of work by Latour on "modes of existence," while turning to reflect on three recent media artworks: James Bridle's *Where the F**k Was I?* (2011), Julian Oliver's *Transparency Grenade* (2012), and the *Post-Cyberwar* (2013) series by Philip Ronnenberg. These projects draw from different strategies to examine how location-aware devices support processes of explication and network curtailment in the production of new cartographies. In this way, such projects deliberately introduce hesitations, cautious trajectories, and uncertain paths in our habitual engagements with locative infrastructure, giving rise to states of deliberation in the inauguration of locative imaginaries.

ACTOR–NETWORK COVERAGE

If locative media can be considered as an engagement with spatial problematics, then what resources has the actor–network supplied in these contexts? More precisely, what has an emphasis on the sociology of associations and a commitment to symmetry enabled in discussions of locative media? Indeed, links with ANT might at first appear intuitive inasmuch as both converge on emergent territories and tagging the nonhuman, along with an emphasis on scalability and "immanent" mapping techniques.[20] However, an argument can be made that the link is equally motivated in confrontation with certain criticisms of utilizing location-aware technologies for creative purposes. For instance, in the well-known essay, "Drifting Through the Grid: Psychogeography and Imperial Infrastructure," Brian Holmes highlighted the problematics for autonomy with reference to plans in the United States for *total information awareness* under the Bush Administration.[21] In his account at the time, liberalized military systems gain their legitimacy and become ideologically reinforced by the participation of a "world citizen." Interpellation is reinvigorated as a technical operation that follows GPS signals from the satellite atomic clock in orbit or triangulation from cell towers to a personal receiver on the ground. These exchanges are based on the promise of security, but to be hailed by this system is to enter into a relation with "Imperial time" (an allusion to Michael Hardt and Antonio Negri's theory of the global society of control). In this sense, the locative media slogan "know your place" takes on a sinister inflection when considered as the integration of civil society with mobilized systems of targeting and surveillance. In line with this account, Andreas Broeckmann infamously labeled early locative media projects as the avant-garde of the society of control.[22]

In many ways, Holmes delivers a familiar Heideggerian refrain in critical perspectives on digital and networked technologies. Holmes's work recalls, for instance, Paul Virilio's provocation that convergence culture is ultimately motivated by a desire for "total vision," where access to global perception also involves an agreement to be controlled in the form of a counterimage.[23] Locative systems, likewise, might be conceived as operational media that stress modeling, patterns over presence, and the agonistic feedback of tracking, targeting, and manoeuvres.[24] As Philip Agre has argued, this model uses predominantly *linguistic* means to actively intervene and reorder contextual behaviors.[25] These are the coordinates that are now amplified through regimes of location awareness, and here the import of Holmes' critique becomes clear because these materialities are enveloped with power by the state authorities and black-boxed as property by corporate services, spreading "manageable" states of nontransparency and opacities of communication for user groups.[26]

Conceptual resources from ANT have, accordingly, been developed against what is perceived as overly reductive assessments of power. Here, discourses on locative media have turned to conceptions of distributed agencies to connect with new dynamics of spatialization, adopting an ethos of heterogeneous relations and multiple ontologies. Uniting these features, the actor–network in particular has become an important concept that offers important reconsiderations of emergent change without a linear gesture to agonistic contestations. It has been used, for instance, to outflank criticisms that locative media devices are necessarily compromised by their imbrication with processes of commodification and the elaboration of military logistics. The idea involves breaking empirically and conceptually with preformatted oppositions of structure and agency, instead working "in between" by accentuating the multiplicity of relations that allows for the circulation of sociotechnical agency.[27]

As an early practitioner and theorist of locative media, Marc Tuters has been an influential figure in establishing this conceptual agenda.[28] His ANT-inspired writings, in particular, have aimed to circumvent assertions that locative arts should engage with how these technologies contribute to the attrition of quotidian experience or the everyday.[29] For Tuters, such "critical" impulses problematically lapse into an obligation for producing Situationist-style interventions, often based on a flawed *mannerist* avant-gardism. The ambition, on the contrary, should be to move beyond this exaggerated and compromised mode of critique and on to a series of more "cosmopolitical" concerns where "artists and designers give voice to mute nonhuman things, each of which can be thought of as having its own ontological reality."[30] Here, Latourian insights, along with related philosophical work on object-oriented ontology, are perceived as a way of reshaping the field to free up new spaces of investigation and testing, rather than projecting forth a domineering structure that replicates a monotonous style of uninventive repetition. At the same time, however, a number of ambiguities are introduced when attributing political capacities to locative projects through the actor–network without also grasping the differential stakes involved in a specifically artistic transduction of space. In Tuters' work, the turn to distributed agencies might discredit asymmetrical and unproductive styles of inquiry, but this new freedom also threatens to dismantle any explanation of the differences established by the projects under consideration. Removing the rough lens of mannerist Situationism seemingly leads to a fundamental dilemma: "the position from which to measure."[31]

ANT-style locative theory, meanwhile, has taken on the tone of a low affirmative address when confronted with the increased economic circulations of value and valorization attached to location-aware technologies. Galloway's sociology of "expectations," for instance—a twist on ANT-derived perspectives that highlights the notable ethical quandaries that accompany locative media projects—places an affirmationist inflection on uncertain modes of sociotechnological becoming under a banner of hope in the future.[32] Her position is also supported by distinctions between analytic frameworks that maintain a consistent vision of space and those that consider spatialities as perpetually performed by interrelations between various entities—a world constituted through individuating differences. While also engaging with the domain of speculative design and experimental arts like Tuters, the coordinates for progressive action are posited to some extent unknown in advance through potential overflows of human and nonhuman action.[33] ANT does not take "human" agencies as necessarily inferior or diminished in agency, Galloway observes, but neither does it presume that empirical realities of an actor–network will resemble states of equality for all entities. Indeed, as her account makes clear, the so-called symmetry of relations is enacted only as a virtual proposition.

In this respect, a central hitch with distributed agencies lies with the connections to apparent internal boundaries through which different domains are formulated. Certainly, for the nonhuman in the context of locative media devices and ensembles, one might consider the layering and obfuscation of software trackers targeting user behavior or the contractual attachments that are by default annulled when jailbreaking, or gaining root access to smartphones. One might additionally think of encounters with the technolegal assemblage of Android as it supports the collection of user metadata for targeted advertising by framing openness through exclusionary standards.[34] How should these arrangements of infrastructure be understood for those practitioners invested in the pursuit of collective environmental change? Indeed, such developments recall an early observation by Marilyn Strathern of a tendency in Latourian perspectives to stress aspects of distributed agencies that are purely characterized by expansion, diffusion, and limitlessness, while neglecting the empirical curtailment of a network in assorted settings.[35] Part of her discussion is a reply of sorts to ANT work that is based on a set of long-term questions around belonging in anthropology, including lines of fidelity, but also instances of property based on the real abstractions of the commodity form. Such trajectories allow stratifications to be run through the actor–network. As Strathern argues, taking account of the key points involving a condensation or truncation in otherwise endlessly expanding networks is crucial. It involves recognizing a moment where "the hybrid object . . . [is] condensed into a single item [of] diverse elements from technology, science and society, enumerated together as invention and available for ownership as property."[36] The implementation of claimants as owners over a network is based on boundaries of belonging. These asymmetries are formed through the legal rights of property and arrangements of governance. Nonhuman and human are mixed, but by delimiting processes that restrain other interests from taking hold. This is the actor–network in manipulable form, a mechanical knot, and a tendency increasingly found across the contemporary production of information space.

In light of these issues, however, it is noteworthy that Latour has recently outlined a renewed approach to professed shortcomings with the actor–network concept through *An Inquiry into Modes of Existence* (*AIME*), a wildly ambitious rethinking of his work that holds some suggestive new paths for theories and practices of locative arts. While appearing in many ways as a synthesis of previous ethnographic studies and taking the form of a digital humanities project, this "anthropology of the Moderns" originates with an appreciation that the actor–network opens up new perspectives but cannot by itself grasp the specificities that characterize the emergence of distinct ontoepistemological domains, such as law, technology, and religion. In other words, there is an explicit acknowledgment that the actor–network needs augmentation in order to accurately render separate and distinct modes of being. However, this still does not operate by eliminating possibilities but rather by elaborating trajectories that unfold via prepositions. For Latour, there is never any level running "underneath" that falls away from these trajectories; there are no "conditions of possibility," only modes that propel possibilities forward along a variety of vectors. There are no "cuts" in the vocabulary of Latour in general. Instead, for the *AIME* project, it becomes a question of how felicity conditions are established that allow for particular "passes," folds, or extensions to occur. Despite always being *on the surface*, as it were, a domain's networked consistency or immanence moves along veiled links in the form of habits. Veiling, likewise, is not a decisive gesture of withdrawal, rejection, or denial, but an act that simply permits an expediency of reproduction. In fact, there is a concerted attempt in *AIME* to revive an affirmative sense of the habitual to the extent that it perpetuates coindividuations, especially in light of revitalizing institutions. Unless a problem is encountered, habits can faithfully replicate themselves efficiently in veiled sequence. However, this is not to say that some relations should not then be reformed; interestingly, the detection of "bad habits" is a task described through the computational analogy of spam (there is a frequent use of digital metaphors throughout) and given over to philosophy as an antispam filter; nevertheless, as I will argue, other trajectories are opened up through the intersection of artistic practice that are also suggestive of how to unbind habitual modes of being for locative media ensembles.

In terms of creative practice, the discussion of esthetics in *AIME* is a key point of interest. Described as *beings of fiction*, these revolve around localized semiologies proper to each mode of existence. A main contention here is that these trajectories are embedded in "the real world," even if they delineate and define domains by the designation of symbolic meaning: "in this inquiry, trajectory, being and direction, sense, or meaning, are synonyms."[37] This issue pertains to how beings of fiction become engaging for subjects, especially by demanding interpretations that unfold a work and subjective experience together into multiple worlds. Here, *subjective* comes to mean processes of subjectivation, especially to the extent that heterogeneous entities become subject to interpretative prepositions:

> If the interpretations of a work diverge so much, it is not at all because the constraints of reality and truth have been "suspended" but because the work must possess many folds, engender many partial subjectivities, and because the more we interpret it the more we unfold the multiplicity of those who love it as well as the multiplicity of what they love in it.[38]

In this case, suspended constraints refer mainly to the idea that beings of fiction are not restricted to some primary bifurcation like the real and the symbolic. Nonetheless, the fact that these experiences develop across a diverse range of trajectories and velocities within heterogeneous modes suggests that beings of fiction are capable of a unique sort of extension. Indeed, while an emphasis is placed on language and positive attachments for Latour, this specific domain essentially reworks a general Kantian resolution of telos and mechanical regularity through the recognition of *purposiveness*. The latter is more or less amended by the general category of figuration that allows for different modes to become meaningful and to "figure out" their realities for themselves. Interestingly, it is described as taking on the main process of *instauration* and is primarily elaborated with reference to the work of artistic creation by Étienne Souriau as a key source for *AIME*.[39]

The notion of beings of fiction relies predominantly on discussions of beauty, language, and narrative but draws some attention to the wider domain of esthetic thought, especially complex dimensions of heteronomy and autonomy. One might, for instance, illustrate this distinction with reference to cases of artistic practice at the intersection of environmental design and politics found in revised accounts of the actor–network by political philosopher Noortje Marres. Here, design objects are situated in spheres of material participation that already act out certain ontoepistemological commitments through the redistribution of problems.[40] Examining devices like issue-mapping tagclouds, sensor-laden ecohomes, and Arduino-based speculative objects, such as augmented teapots for conscious energy usage, these entities are seen to draw together both signifying and asignifying dimensions in a mode of emergent political action. In their own way, they can also be listed as spatiotemporalizing technologies, as devices that carry forth their own sense and perception of location. Interestingly, the work of locative artist Esther Polak is also included as a point of distinction in Marres' discussion, especially the piece *Spiral Drawing Sunrise* (2009), a sort of sundial on wheels that utilizes sunlight to trace out recordings of the sun's path in public space. This robotic device marks out the sunrise with sand through slow spirals, delineating moments that also register the passing of traffic and people and shadows cast by nearby buildings. As a media art project, the work is located across a number of different genres of practice and additionally suggests an ambiguous set of questions pertaining to collective habitual action. For Marres, it *dramatizes* the problem of relevance concerning environmental issues but ultimately "does not count as a solution to the predicament of environmental participation: it does not tell us how the mutual relevance of issues, actors and spaces of participation can be ensured."[41]

What should be made of experiential domains that append an impression of the problematic without resolution? Indeed, a prolongation of sociotechnical problems can be seen as recurring throughout particular locative art projects as an expression of an antipositivist technicity. This often appears as a mode of questioning that works to drive veiled trajectories or solutions back to the purported problems that they were aimed to address. In doing so, what Latour describes as beings of fiction are utilized to interrupt the procession of habits. Purposiveness provides leverage for the refiguration of relations. Within this context, consider James Bridle's *Where the F*ck Was I?* (2011), a book of data produced from tracking software on his iPhone operating system. The work is broadly situated in his theorization of the "New Aesthetic" as a case of "machine vision" since, rather than providing a human-centered cartography, various points of interest are highlighted by relations between the device, mobile cell towers, and wireless networks.

> This digital memory sits somewhere between experience and non-experience; it is also an approximation; it is also a lie. These location records do not show where I was, but an approximation based on the device's own idea of place, its own way of seeing. They cross-reference me with digital infrastructure, with cell towers and wireless networks, with points created by others in its database. Where I correlate location with physical landmarks, friends and personal experiences, the algorithms latch onto invisible, virtual spaces, and the extant memories of strangers.[42]

Controversially, the consolidated.db file utilized by Bridle for this project in iOS was installed unknown to users for purposes of tracking Hertzean space. Although not specifically designed to target individual trajectories per se, Apple mobilized user populations as free labor to provide analytics on wireless capacities and mobile infrastructures. In his data diary, Bridle demonstrates how these interrelations complicate the location of an autonomous subject in the process. Although this might be taken as further evidence that "we have never been modern," *Where the F*ck Was I?* is notable as a composition of these hybridities that maintain their ambiguous predilections toward curtailment in a state of tension. This is a mechanical knot refigured.

In Latour's *AIME*, esthetics is provocatively described as being an impression of infinitude that comes from the intensity of "vibrations" between undetachable aspects of a work (material, embodied, rooted) and detachable characteristics (selected, arranged, chosen). Despite being illustrated through narrative and literature, Latour's definition is suggestive for considering how such intensities between form and content feed back onto habitual patterns with locative systems. However, this crossing suggests some contrast with predominant considerations of esthetics as pleasurable beauty in *AIME*, inasmuch as these intensities might also convey unease. The recent piece *Transparency Grenade* (2012) by Julian Oliver is a speculative location-aware object that channels such oscillations in the mode of security anxieties around volatile informational extensions or leaks. Presented as a Soviet F1 hand grenade equipped with a microphone, wireless antenna, and computer, the device captures information in an immediate environment when the pin is pulled within a small radius and then transmits this data to a server for purposes of analysis. E-mails, images, website fragments, and audio are then extracted and published on an online map revealing the "detonation" site of the grenade. The work, in this way, "seeks to capture these important tensions in an iconic, handheld package while simultaneously opening up a conversation about just how much implicit trust we place in network infrastructure."[43] The object is designed to enroll a series of actors into a space where relations across informational infrastructures can be reimagined and reformulated in turn. *The Transparency Grenade* is also not a "solution" or a habitual thing but a dramatization of embedded and highly distributed problematics. Fictional beings can be extended in reflective modes across diverse sociotechnical and infrastructural contexts, especially by intensifying subjective experiences that accompany ongoing processes of informational expropriation and enclosure.

While curtailment is a foremost characteristic of locative media, especially through the imposition of limits on the potentials of digital and networked technologies, important strains of locative art have engaged with the alternative relations with these asymmetrical settings. During a period of ongoing controversies regarding the range and intensity of data acquisition, the emergence of new corporate and state centers of calculation and the routine imposition of Internet censorship, there is an urgent need to inaugurate new imaginaries concerning our autonomous access to communication. Philipp Ronnenberg's speculative series *Post-Cyberwar* (2013) provides a final illustrative example of the stakes involved in terms of network "kill switches" and the ideal of free systems. Citing instances during the Arab Spring in Tunisia, Egypt, and Libya, where Internet access was cut off in order to disrupt self-organized protest movements, the trilogy of works by Ronnenberg proposes different methods to address the problem of network censorship, government intervention, and downtime. *OpenPositioningSystem*, for instance, is a media ecological navigation system that recommends harnessing activity produced by generators in power plants, pumping station turbines, and other large-scale industrial machinery to triangulate positions via seismic waves through the ground. The *Teletext Social Network*, meanwhile, utilizes recently abandoned commercially obsolete analogue television broadcasting bandwidth to set up wireless infrastructure maintained by users in a media archeological gesture. Finally, *The Sewer Cloud* is a living and self-reproducing data network in the sewerage system of London based on a method of inserting into and extracting data from the algae species *Anabaena* (a cyanobacterium) in a unique mix of bioart and locative media. Although these projects might perhaps appear like the wild beings of fiction, they are explicitly composed to inaugurate conversations about a collective problem and to support imaginaries through aporetic instauration. In this case, the attachments and desires that crisscross the *Post-Cyberwar* series also seemingly raise "impossible" questions through central tensions of heteronomy and autonomy, particularly as these are manifested along the junction of subjectivation and habits, thereby multiplying worlds by rearranging our relations with other existing infrastructures, devices, and living beings.

Rather than the actor–network concept, I have been arguing throughout this chapter that the significance of these works might usefully be considered through some reflection on dynamics of delineation and esthetics, especially when examining cases of critical technical practice. My approach here has utilized some aspects of Latour's recent *AIME* project to demonstrate the importance of these links but also signal some apparent disjunctures in that vast account of metaphysics. To be clear, on a number of competing registers, my argument has been for esthetics not on the terms of solutions but in relation to problems; to this extent, we might additionally approach any and all esthetic departures with some degree of care, precisely because this spot appears to turn on a multiplicity of vectors, trajectories, or paths in our habitual settings. Esthetics convey responsibilities in an attachment to a potential but, most crucially, in how sociopolitical investments are sensed and perceived in confrontation with problems for the common.

NOTES

1. Tristan Thielmann, "Locative Media and Mediated Localities: An Introduction to Media Geography," *Aether: The Journal of Media Geography* 5A (2010): 1–17.
2. Mike Crang and Stephen Graham, "Sentient Cities: Ambient Intelligence and the Politics of Urban Space," *Information, Communication and Society* 10, no. 6 (2007): 789–817.
3. Matthew Fuller and Sónia Matos, "Feral Computing: From Ubiquitous Calculation to Wild Interaction," *Fibreculture Journal* 19 (2011): 144–163.
4. N. Katherine Hayles, "RFID: Human Agency and Meaning in Information-Intensive Environments," *Theory, Culture & Society* 26, no. 2–3 (2009): 49.
5. Eric Gordon and Adriana de Souza e Silva, *Net Locality: Why Location Matters in a Networked World* (Malden, MA: Wiley, 2011), 4.
6. Andrea Zeffiro, "A Location of One's Own: A Genealogy of Locative Media," *Convergence: The International Journal into New Media Technologies* 18, no. 2 (2012): 249–266.

7. Drew Hemment, "Locative Arts," *Leonardo* 39, no. 4 (2006): 351.
8. Marc Tuters and Kayzs Varnelis, "Beyond Locative Media: Giving Shape to the Internet of Things," *Leonardo* 39, no. 4 (2006): 357–363.
9. Conor McGarrigle, "The Construction of Locative Situations: Locative Media and the Situationist International, Recuperation or Redux?" *Digital Creativity* 21, no. 1 (2010): 55–62; Marc Tuters, "From Mannerist Situationism to Situated Media," *Convergence* 18 (2012): 267–282.
10. Anne Galloway and Matt Ward, "Locative Media as Socializing and Spatialising Practice: Learning from Archaeology," *Leonardo Electronic Almanac* 14, no. 3 (2006); accessed November 18, 2013, http://leoalmanac.org/journal/Vol_14/lea_v14_n03–04/gallowayward.asp.
11. Jeremy Hight, "Locative Dissent," *Sarai Reader 06: Turbulence* (2006), accessed November 18, 2013, www.sarai.net/publications/readers/06-turbulence: 129.
12. Ben Russell, "Know Your Place: Headmap Location Aware Devices," *Headmap Redux*, accessed November 18, 2013, www.technoccult.com/library/headmap.pdf.
13. Bonnie DeVarco, "Earth as a Lens: Global Collaboration and GeoCommunication," in *The TCM Reader* (2004), accessed November 18, 2013, http://rixc.lv/reader/txt/txt.php?id=178&l=en&raw=1.
14. Ulises Ali Mejias, *Off the Network: Disrupting the Digital World* (Minneapolis: University of Minnesota Press, 2013).
15. Carlos Barreneche, "The Order of Places: Code, Ontology and Visibility in Locative Media," *Computational Culture* 2 (2012), accessed November 18, 2013, http://computationalculture.net/article/order_of_places.
16. Martin Dodge, Rob Kitchin, and Chris Perkins, eds., *Rethinking Maps* (London: Routledge, 2009); Martin Dodge and Rob Kitchin, *Code/Space: Software Space and Society* (Cambridge, MA: MIT Press, 2011); Adrian McKenzie, *Wirelessness: Radical Empiricism in Network Culture* (Cambridge, MA: MIT Press, 2010); Nigel Thrift, *Non-Representational Theory: Space, Politics, Affect* (London: Routledge, 2007).
17. Roger Clark, "Information Technology and Dataveillance," *Communications of the ACM* 31, no. 5 (1988): 498–512; Philip E. Agre, "Surveillance and Capture: Two Models of Privacy," in *The New Media Reader*, eds. Noah Wardrip-Fruin and Nick Montfort (Cambridge, MA: MIT Press, 2003), 737–760; Dodge and Kitchin, *Code/Space*, 81–110.
18. Michael Dieter, "The Virtues of Critical Technical Practice," *Differences: A Journal of Feminist Cultural Studies* [forthcoming].
19. Minna Tarkka, "The Labors of Location: Acting in the Pervasive Media Space," in *The Wireless Spectrum: The Politics, Practices and Poetics of Mobile Media*, eds. Barbara Crow, Michael Longford, and Kim Sawchuck (Toronto: University of Toronto Press, 2010), 133.
20. Thielmann, "Locative Media and Mediated Localities," 1–17.
21. Brian Holmes, "Drifting Through the Grid: Psychogeography and Imperial Infrastructure," *Springerin* (March 2004), accessed November 18, 2013, www.springerin.at/dyn/heft_text.php?textid=1523&lang=en.
22. Andreas Broeckmann, "locative? tracked!" in *Exhibiting Locative Media: (CRUMB Discussion Postings)* (May 20, 2004), accessed November 18, 2013, www.metamute.org/editorial/articles/exhibiting-locative-media-crumb-discussion-postings.
23. Paul Virilio, "The Visual Crash," in *CRTL [SPACE]: Rhetorics of Surveillance from Bentham to Big Brother*, eds. Ursula Frohne Levin and Peter Weibel (London: ZKM/MIT Press, 2002), 110.
24. Julian Crandall, "Operational Media," *C-Theory* (2005), accessed November 18, 2013, www.ctheory.net/articles.aspx?id=441.

25. Agre, “Surveillance and Capture.”
26. Philipp von Hilgers, “The History of the Blackbox: The Clash of a Thing and Its Concept,” *Cultural Politics* 7, no. 1 (2011): 41–58.
27. Bruno Latour, *Reassembling the Social: An Introduction to Actor–Network-Theory* (Oxford: Oxford University Press, 2005).
28. Tuters and Varnelis, “Beyond Locative Media.”
29. Marc Tuters, “From Control Society to Parliament of Things: Designing Object Relations with an Internet of Things” (paper presented at the Digital Arts and Culture Conference, University of California, Irvine, December 12–15, 2009), accessed November 18, 2013, http://escholarship.org/uc/item/3zj2t89z#page-8.
30. Tuters, “From Mannerist Situationism to Situated Media,” *Convergence: The International Journal of Research into New Media Technologies* 18, no. 3 (2012), 271.
31. Tuters, “From Mannerist Situationism to Situated Media,” 275.
32. Anne Galloway, “Locating Media Futures in the Present, or How to Map Emergent Associations and Expectations,” Aether: The Journal of Media Geography 5A (2010): 27–36; Anne Galloway, A Brief History of the Future of Urban Computing and Locative Media, PhD Thesis (Ottawa: Carleton University, 2008).
33. Anne Galloway, “Emergent Media Technologies, Speculation, Expectation, and Human/Nonhuman Relations,” *Journal of Broadcasting & Electronic Media* 57, no. 1 (2013): 53–65.
34. Kimberley Spreeuwenberg and Thomas Poell, “Android and the Political Economy of the Mobile Internet: A Renewal of Open Source Critique,” *First Monday* 17, nos. 7–2 (2012), accessed November 19, 2013, http://firstmonday.org/ojs/index.php/fm/article/view/4050/3271.
35. Marilyn Strathern, “Cutting the Network,” *The Journal of the Royal Anthropological Institute* 2, no. 3 (1996): 517–535.
36. Strathern, “Cutting the Network”: 525.
37. Bruno Latour, *An Inquiry into Modes of Existence: An Anthropology of the Moderns*, trans. Catherine Porter, Cambridge, MA: Harvard University Press, 2013, 237.
38. Latour, *An Inquiry into Modes of Existence*, 243.
39. Bruno Latour, “Reflections on Etienne Souriau’s Les différents modes d’existence,” in *The Speculative Turn: Continental Philosophy and New Materialism*, eds. Levi Bryant, Nick Srnicek, and Graham Harman (Melbourne: re.press, 2011), 310–311.
40. Noortje Marres, *Material Participation: Technology, the Environment and Everyday Publics* (London: Palgrave MacMillan, 2012).
41. Marres, *Material Participation*, 155.
42. http://booktwo.org/notebook/where-the-f-k-was-i/
43. Susanne Staunch, *Transparency Grenade* (2012), accessed November 18, 2013, http://transparencygrenade.com/.

Contributors

César Albarrán-Torres is a doctoral researcher at the Digital Cultures Program, University of Sydney. He has worked extensively in academic and nonacademic publications as an author, editor, film critic, and translator. His current research delves into the cultures that form around the digitization of gambling in online casinos, mobile apps, and electronic gaming machines. Other research interests include cross-platform television and film narratives, download culture, and the construction of political personas in online realms. He has written chapters in three books on Mexican media: *Reality shows un instante de fama* (2003), *Internet: Columna vertebral de la sociedad de la información* (2005), and *Reflexiones sobre cine mexicano contemporáneo* (2012).

Mark Andrejevic is Deputy Director of the Centre for Critical and Cultural Studies, University of Queensland, and the author of *Reality TV: The Work of Being Watched*, *iSpy: Surveillance and Power in the Interactive Era*, and *Infoglut: How Too Much Information Is Changing the Way We Think and Know*, as well as numerous articles and book chapters on surveillance, popular culture, and digital media.

Carlos Barreneche is an Assistant Professor at Universidad Javeriana (Colombia) Communication Department. He received his PhD at CAMRI, University of Westminster (UK), in 2012. He also received degrees in philosophy and psychology. His research areas cover the political economy of digital media, biopolitics of network culture, software studies, and media theory. His articles have appeared in journals such as *Convergence: The International Journal of Research into New Media Technologies*, *Computational Culture*, and *Media, Culture & Society*.

Peter Bayliss is a Lecturer in the Faculty of Health, Arts and Design, Swinburne University of Technology, where he teaches and researches in the fields of games and interactive media and media studies. Peter recently completed his PhD research at RMIT University's School of Media and Communication. His thesis focused on the central importance of the player's embodied being for their phenomenological experience of videogame play. His research interests include phenomenology and interactive media, mobile media, novel/emerging interface modalities and experimental game design, and the history of videogame cultures and industries.

Michael Dieter is a Lecturer in new media at the University of Amsterdam. His research is focused on relations among media esthetics, ecological thought, and political philosophy in critical technoscientific art practices, and his articles have appeared in the journals *M/C*, *Fibreculture*, and the *Australian Humanities Review*.

Timothy Dwyer is Senior Lecturer in the Department of Media and Communications at the University of Sydney, Australia. He teaches media law and ethics to undergraduate and postgraduate students, and his research focuses on the critical evaluation of media and communications industries, regulation, media ethics and policy. He is the author of *Legal and Ethical Issues in the Media* (Palgrave Macmillan, 2012) and *Media Convergence* (Open University, 2010), and he is the coeditor (with Virginia Nightingale) of *New Media Worlds: Challenges for Convergence* (Oxford, 2007). Before moving to academia in 2002, he has worked for the Australian Broadcasting Corporation (1981–1989) and the Australian federal government agencies responsible for privacy rights (1990–1994) and for electronic media regulation.

Jason Farman is an Assistant Professor at the University of Maryland, College Park, in the Department of American Studies, a Distinguished Faculty Fellow in the Digital Cultures and Creativity Program, and a faculty member with the Human–Computer Interaction Lab. He is author of the book *Mobile Interface Theory: Embodied Space and Locative Media* (Routledge, 2012; winner of the 2012 Book of the Year Award from the Association of Internet Researchers), which focuses on how the worldwide adoption of mobile technologies is causing a reexamination of the core ideas about what it means to live our everyday lives: the practice of embodied space. His second book is an edited collection titled *The Mobile Story: Narrative Practices with Locative Technologies* (Routledge, 2013). He is currently working on a book project called *Technologies of Disconnection: A History of Mobile Media and Social Intimacy*. He received his PhD in digital media and performance studies from the University of California, Los Angeles.

Gerard Goggin is ARC Future Fellow and Professor of Media and Communications at the University of Sydney. His research interests lie in mobiles, the Internet, and new media; media policy and regulation; and disability. Gerard's books include *Routledge Companion to Mobile Media* with Larissa Hjorth (2014); *Disability and the Media*, with Katie Ellis (2014);

New Technologies and the Media (2011); *Global Mobile Media* (2011); *Cell Phone Culture* (2006); and, with Christopher Newell, *Disability in Australia* (2005) and *Digital Disability* (2003). His edited collections on mobiles include *Locative Media* (2014) and *Mobile Technology and Place* (2012), both with Rowan Wilken; with Larissa Hjorth, *Mobile Technologies: From Telecommunications to Media* (2009); and *Mobile Phone Cultures* (2008). Gerard has also published three edited volumes on the Internet: with Mark McLelland, *Routledge Companion to Global Internet Histories* (2015); *Internationalizing Internet: Beyond Anglophone Paradigms* (2009); and *Virtual Nation: The Internet in Australia* (2004).

Larissa Hjorth is an artist, digital ethnographer, and Professor in the Games Programs, School of Media & Communication, RMIT University. She is codirector, with Heather Horst, of RMIT's Digital Ethnography Research Centre (DERC). Since 2000, she has been researching the gendered and sociocultural dimensions of mobile, social, locative and gaming cultures in the Asia–Pacific; these studies are outlined in her books, *Mobile Media in the Asia–Pacific* (Routledge, 2009); *Games & Gaming* (Berg, 2010); with Michael Arnold, *Online@AsiaPacific: Mobile, Social and Locative in the Asia–Pacific Region* (Routledge, 2013); with Sam Hinton, *Understanding Digital Media in the Age of Social Networking* (Sage, 2013); with Ingrid Richardson, *Social, Locative and Mobile Gaming* (Palgrave, 2014). Larissa has coedited four Routledge anthologies: with Dean Chan, *Gaming Cultures and Place in the Asia–Pacific Region* (2009); with Gerard Goggin, *Mobile Technologies: From Telecommunication to Media* (2009); with Jean Burgess and Ingrid Richardson, *Studying the iPhone: Cultural Technologies, Mobile Communication, and the iPhone* (2012); and. with Katie Cumiskey, *Mobile Media Practices, Presence and Politics: The Challenge of Being Seamlessly Mobile* (2013).

Nadav Hochman is a doctoral candidate in the History of Art and Architecture Department at the University of Pittsburgh and a visiting scholar in the Software Studies Initiative at the Graduate Center, CUNY. His current research focuses on the use of computational methods for the analysis of massive online visual cultural data sets. Using data visualization techniques from a digital humanities perspective, Nadav's work examines how and what we can learn about local and global cultural patterns and trends by aggregating large amounts of user-generated visual materials. He holds master's degrees from the University of Pittsburgh, as well as from the Interdisciplinary Program of the Arts at Tel-Aviv University.

Tama Leaver is a Lecturer in the Department of Internet Studies at Curtin University in Perth, Western Australia, and a research fellow in the ARC Centre for Excellence in Creative Industries and Innovation working in Curtin's Centre for Culture and Technology. His research interests include digital death, social media, online identity, and media distribution. He has published in a number of journals, including *Popular Communication*, *Media International Australia*, *Comparative Literature Studies*, and the *Fibreculture* journal. He is the author of *Artificial Culture: Identity, Technology and Bodies* (Routledge, 2012) and coeditor of *An Education in Facebook? Higher Education and the World's Largest Social Network* (Routledge, 2014). Tama is @tamaleaver on Twitter, and his main web presence is www.tamaleaver.net.

Clare Lloyd is from the Department of Internet Studies at Curtin University. She specializes in mobile communication and mobile media. Her recent publications include the coauthored papers "Consuming Apps: The Australian Woman's Slow Appetite for Apps" (2012) and "Fun and Useful Apps: Female Identity Construction and Social Connectedness Using the MobilePhone Apps" (2012).

James Meese is a doctoral student at the Institute for Social Research, Swinburne University of Technology in Melbourne, Australia. His research has been published in *Television and New Media*, *The European Journal of Cultural Studies*, and *Computers and Composition*. He is currently completing a dissertation examining how formal and informal cultures of copyright law interact to constitute legal subjects. He also conducts research on informal economies, postbroadcast television, intellectual property, information policy, and media sport.

Didem Özkul is a PhD candidate at the Communications and Media Research Institute (CAMRI) at the University of Westminster, London, having started the PhD program at CAMRI in September 2010. Before starting her PhD, she received her MA degree in media and visual studies at Bilkent University, Turkey, where she analysed image–spectator interaction in immersive digital platforms. Her current research concerns people's use of mobile communication technologies in augmented urban spaces, particularly focusing on spatial perception. For her research, she engages in creative methodologies (mental maps) in order to explore and visualize the transformations in sense of place and whether mobile media has an affect on those transformations (www.mobilenodes.co.uk). Currently, she is also working as a research associate at the University of Westminster for a project funded by the Arts and Humanities Research Council entitled 'Community-Powered Transformations' (www.digitaltransformations.org.uk), a research network exploring digital transformations in the creative relationships between cultural and media organizations and their users.

Ned Rossiter is Professor of Communication in the School of Humanities and Communication Arts at the University of Western Sydney, where he is also a member of the Institute for Culture and Society. He is currently working on two books entitled *Logistical Nightmares: Infrastructure,*

Software, Labour (Routledge, forthcoming, 2014) and, with Geert Lovink, *Urgent Aphorisms: The Politics of Network Cultures* (Minor Compositions, forthcoming, 2014).

Raz Schwartz is a postdoctoral researcher at Cornell Tech New York City. Raz studies social media usage in urban settings and focuses on examining local social interactions by applying computational social science methods. Prior to joining Cornell Tech NYC, he was a visiting scholar at the Social Media Information Lab at Rutgers University. Raz completed his PhD in the STS program at Bar-Ilan University and was a visiting scholar in the Human Computer Interaction Institute at Carnegie Mellon University. Raz holds a master's degree in law from Bar-Ilan University and a bachelor's degree in communications and economics from Tel-Aviv University. His research on location-based social networks has been presented in various academic settings and featured in media outlets such as the *Wall Street Journal*, *Wired*, *Rhizome*, and *The Atlantic.*

Federica Timeto holds a PhD in Aesthetics of New Media at the University of Plymouth, UK, and is currently completing her PhD in sociology at the University of Urbino Carlo Bo. Her scholarship encompasses studies of locative media, visual and cultural studies, feminist aesthetics, and epistemology. She edited *Culture della differenza* (Utet, 2008), a reader on the intersection of feminism, visuality, and postcolonial issues.

Rowan Wilken is Senior Lecturer in Media and Communications and holds an Australian Research Council–funded Discovery Early Career Researcher Award (DECRA) in the Swinburne Institute for Social Research, Swinburne University of Technology, Melbourne, Australia. His present research interests include locative and mobile media, digital technologies and culture, domestic media consumption, old and new media, and theories and practices of everyday life. He is author of *Teletechnologies, Place and Community* (Routledge, 2011) and coeditor, with Gerard Goggin, of *Mobile Technology and Place* (Routledge, 2012).

Andrea Zeffiro is a researcher and writer whose work intersects the cultural politics and practices of emerging technologies, contemporary media histories, feminist media studies, and multidisciplinary research methods. Over the last ten years, Zeffiro has worked as an ethnographer in a number of transdisciplinary research formations alongside artists, designers, social scientists, computer scientists, engineers, and medical doctors. Prior to her academic pursuits, Zeffiro spent a number of years drafting and implementing garment-purchasing policies for the public sector while channeling her creative energies toward AMBUSH: a line of clothing designed and created from second-hand garments. Zeffiro is Adjunct Professor in the Department of Communication, Popular Culture and Film, at Brock University, St. Catharines, Ontario, Canada.

Index

ABC vs. Lenah Game Meats 142–3
absent presence 33
activism: activist tech 8; locative media and 13, 66–78
actor–network theory (ANT) 10, 12, 101, 226–7; coverage 227–34; locative esthetics and 224–34; locative style 229; *see also* locative media art
advertising, *see* promotional culture
aesthetic: as dimension of creativity 96; of disappearance 163
affective computing 199–201
Africa: locative media in 1
agentive capacity 202
Agre, Philip 228
Ahmed, Sara 87, 88, 89–90
AIME project 230–1, 233, 234
Albrechtslund, Anders 166
Alcocer, Alejandro González 153
Allan, Alasdair 169
Amazon 122
Anderson, Chris 203
Andreas, Peter 75
Android platforms 6; *see also* smartphone
Angry Birds 1
AOL 126
Apple 163, 169–70, 171: App Store 164; iOS4 location data log 169; 'It Just Works' slogan 162, 165; Macintosh 163, 164; privacy policies 164–5
Apple Maps 9, 83, 91; launch glitches 83–4; see also maps
application programming interface (API) 127
apps 6; industry in Mexico 150; information-gathering capacity 123–4, 127; ubiquity of 164
AT&T: utilization of user data 193, 194
Australian Communications and Consumer Action Network (ACCAN) 123
Australian Communications and Media Authority (ACMA) 128, 143
Australian Law Reform Commission (ALRC): inquiry into locative media 142

Barreneche, Carlos 137
Barkhuus, Louise 63
Bartholl, Aram 100; *15 Seconds of Fame* 100
Baudrillard, Jean 87
Bauman, Zygmunt 133
Beck, Glen 66
Bennett, Jane 202, 203, 204
big data 2, 121, 184, 196, 203
Bing 122
Blades, Mark 41
Blanchot, Maurice 7
Blast Theory 4
blogs: microblogs 26
body–technology relations, *see* human–object relationship
Bogost, Ian 13, 202–3, 204
Bolter, Jay David 89
Bonacich, Edna 212
border: poetics 71–5l; spaces 75–7
boyd, danah 35
Brewer, Jan 70
Bridle, James 10, 227, 232; *Where the F••k Was I?* 227, 232
Brown, Wendy 75

Calderón, Felipe 151, 153, 157
camera phone 4, 23–35; changing practices 25–7; creation of

personal data archive 34; distribution of images 26–7; emplacement of images 26–7; gendered practices 35; self-portraiture 25; sharing via 29–30
Cárdenas, Micha 66
Carey, James 217
Carroll, Amy 66
cartographies: mediated 23; emplaced 25; representational 99; critical 99; postrepresentational 100; *see also* maps
Casey, Edward 88
Cell-All 197, 198–9, 201
cellular mobile networks 4; *see also* mobile phones; smartphones
children: in-app purchasing 123; social media issues concerning 123
Children's Online Privacy Protection Act (US) 123
China: locative media in 1
city: automatic production of 112–15; diagramming of 112–13; mobile media and 11; remediation of 113
Clarke, Arthur C. 163
cloud management 123
code/space 11, 216
Cohen, Julie 165
containerisation 211–12; *see also* logistical media
Contingente MX blog 154, 156
convergence culture 227
cookies 128–9
Cordero, Olga Sánchez 157
Cowen, Deborah 212
Craib, Raymond 90
Cramer, Henriette 63
Cranshaw, Justin 54
Crawford, Kate 169
Creepy 163, 170
Crowley, Dennis 113, 179, 180–1, 182, 184, 185, 187; marriage 181; public profile 181; *see also* Foursquare
Cyworld 6; Egloo 34; minihompy 27, 29, 32, 33

data, *see* user data
dataveillance 224; *see also* locative surveillance; user data
De Souza e Silva, Adriana 3, 5, 11, 101, 164, 165, 166, 167, 225
Deleuze, Giles 9, 10, 108–9
DeNicola, Lane 162, 163–5
Deseriis, Marco 72–3
DeVarco, Bonnie 225
Díaz, José Ramón Cossío 157
Dixon, David 218
do not track (DNT) 127–30; *see also* opt-out policies
Dodge, Martin 11, 13, 87, 100, 216
Dodgeball 2, 4, 179, 182
Dominguez, Ricardo 66, 67, 72, 73, 74
drones 1; drone divide 205; drone thinking 203–4; place in popular imagination 195–7; portable locative media devices as 11, 193–205; smartphones as, 197, 198–9; as surveillance device 196; theory of 201–5; uses 196
Dropbox 180

eBay 122
Ek, Richard 96, 100–1
emplacement 26–7
end-user license agreements (EULAs) 164
Engemann, Christoph 209
Enterprise Resource Planning (ERP) systems 208, 209, 213; *see also* SAP
everyday practice 7–10
expectations, sociology of 229

Facebook 24, 68, 109, 121, 126, 150, 167; privacy issues 122–3
Facebook Places 2, 112, 123, 125
Facebook Poke 123
Farman, Jason 85
FBI Office of Cybercrimes 66
Field Trip 137
FinFisher 156
Finland: locative media in 1
Fix, E. 109–10
flat ontology 13
Flickr 6, 123, 170
Foley, Michael 7
Ford, Sam 167–8
Foster Holleman, Hannah 126
Foster, John B. 126
Foucault, Michel 10
Foursquare 2, 11, 13, 25, 35, 55, 88, 112, 113, 123, 125, 131, 148, 150, 166, 167, 168, 171, 177–8, 226; Agora 183; ARstreets 183;

corporate strategy 178; cross-platform partnerships 184; development of business model 179–82; engagement with location 179; Explore 185; financial returns 181; FourGraph 183; Foursquare for Business 186; FourSquare Lists 183; Graph Search 185; long-term sustainability 184–5; monetisation strategy 185; non-Anglophone services 179; open API 182–4; political economics 177–87; reliance on user base 178, 183–4; Sonar 183; When Should I Visit 183; World of Fourcraft 183; *see also* Crowley, Dennis
Frith, Jordan 3, 11, 164, 165, 166
Fuller, Buckminster 225
Fullilove, Mindy T. 45

Galloway, Anne 225, 229
gaming: in-app purchasing 123; *see also* apps
Gates, Bill 198
Gaydar 4
generative rules 115
geodemographic information systems (GDIS) 107, 108; cluster diagrams, 107–16; clustering classification 108–10; geodemographic bias 108; geodemographic logic 108; NN rule 109–11, 113, 115
geodemographic spatial rationality 11
geographic information systems (GIS) software 5, 95, 225
geography: locative media and 5
Geolocalization Law (Mexico) 14, 149–50, 153–7; opposition to 154–5; ratification 157
geolocation services: democratisation 6
geotagging 23, 34, 35, 166; exclusions 62–3; public park study using 53, 56–62; tracing experiences using 59–62
Gerlitz, Carolin 179
Gillespie, Tarleton 13
global society of control 227
Goggin, Gerard 6, 164, 169
Golding, Peter 178
Google 121, 122, 126; dominance of mobile advertising market 127
Google Earth 5; privacy issues 122, 130
Google Glass 1, 14, 171; heads-up display (HUD) 136; locative media apps 136–7; privacy issues 130, 136–45; public reaction 138
Google Latitude 179
Google Maps 5, 41, 43–5, 88, 111, 123, 130, 224; micronavigation 49; removal from iPhone 5 83; Street View 45–6, 122, 138–9; *see also* maps
Google Places 179
Google+ 123
Gordon, Eric 5, 101, 167, 225
Gowalla 182
GPS 1, 4, 68, 122
Green, Joshua 167–8
Greenleaf, Graham 142
Grindr 1
Grusin, Richard 89
Guattari, Félix 96
Guía Pemex app 150

Hampton, Keith N. 55
Hansen, Lone Koefoed 91
Hansen, Mark 87
Haraway, Donna 86
Hardt, Michael 227
Harris, Richard 109
Hartley, John 167
Hayles, N. Katherine 179, 224
Heidegger, Martin 227
Helmond, Anne 179
Hepworth, Katie 212
Hernández, Sergio Valls 157
Hewlett, Robin 100; *Street with a View* 100
Hight, Jeremy 225
Hjorth, Larissa 63, 164, 165
Hochman, Nadav 55
Hodges, J. L. 109–10
Holmes, Brian 227
Homeland Security Advanced Research Projects Agency (HSARPA) 199
homophily principle 114–15
human–object relationship 7, 12, 202, 224; *see also* drones
Humphreys, Lee 63
Hunter Punta Tracking 156
Husserl, Edmund 87

I Can Stalk U 163, 168–9, 170
I Love Coffee 24, 28

illegal immigrants from Mexico 70–1; *see also* border poetics; Transborder Immigrant Tool
Immigration Policy Institute 70
Ingold, Tim 27
Innes, Harold 84
Instagram 23–4, 25, 26, 53, 55–6, 123, 148, 187; metadata 56
Internet of Things (IoT) 6, 218, 219, 224
iPad 44, 163
iPhone 5, 6, 163, 164, 224; collection of information 123; *see also* smartphone
iPhone Tracker 163, 169, 170
Ito, Mizuko 23, 30, 32
Iusacell 150, 156
Ive, Jonathan 163

Jenkins, Henry 167–8
Jiepang 1, 25, 35

Kakao 1, 23, 27–9; apps 24; image-sharing 29–35
KakaoGame 29
KakaoPlace 25
KakaoStory 29, 30–2, 33, 34
KakaoTalk 28, 30
Kakavas, Ionnis 170
Kalnins, Karlis 3, 225
Kanngieser, Anja 210
Karasic, Carmin 72
Keane, John 129
Kenya: locative media use 148; *see also* Ushahidi
Key Performance Indicators (KPI) 212
Kim, Beom-Soo 28
Kindle 44
Kinsley, Ben 100; *Street with a View* 100
Kitchin, Rob 11, 13, 41, 87, 100, 216
Korea: locative media in 23–35; PC bangs 29
Koskinen, Ilpo 23

La Ley de Geolocalización, see Geolocalization Law
Landau, Susan 193
landmarks, 41–2, 50; *see also* maps; spatial experiences
Lapenta, Fracesco 27
Lash, Scott 115
Latour, Bruno 10, 102, 226–7, 228, 229, 230–1, 234
Law, John 110
Lefebvre, Henri 7, 69–70, 76, 85, 114
Levine, Paula 90
LinkedIn 109, 121, 123
liquid modernity 133
Livehoods project 54
location: anxiety 225; awareness of 225; concept of 2–4;
location-aware technologies 95
location-based services (LBS) 2, 23–29, 35, 121; mainstreaming on smartphones 35; privacy issues 130–2; use by Korean women 23–5
locative media art 4, 10, 94, 96, 226; *see also* performative mapping
locative media: academic research 7; efficacy of term 3, 68–9; exceeding of location 100–102; experience 95; history 4–6; material aspects 95; symbolic aspects 95; technologies and infrastructure 3–4; in urban design 4; *see also* activism; locative media art; locative networking; logistical media
locative networking 107–16; *see also* geodemographic information systems (GDIS)
locative praxis 8, 69–70, 77
locative surveillance 194, 224; types of 166–7; collateral 166; corporate 166; government 166; social 166; user awareness tools 167–70; mood monitoring 199; *see also* drones
Lofland, Lyn H. 63
logistical media: containerisation and 211–12; location of 210–12; locative media as 12, 208–19; purpose of 211; soft infrastructure 215–17; software 212; theory 217–19
London: commuting culture 44; imageability of city 42; spatial navigation 9, 39–50; spatial practices 9; *see also* landmarks; maps; Streetmuseum project
Loopt 123
Lorimer, Hayden 9
Lovegety 4

Lukermann, Fred 2–3
Lynch, Kevin 40

Mackenzie, Adrian 97, 100
Manovich, Lev 55, 163
Mansell, Robin 178
maps 23; cognitive 40–2; embodied 85–7, 90; historical 86–7; in locative media projects 88; mobile 40–50, 83; power dynamics 90; as representational objects 85; as route to discovery 47–9; satellite visualization 87; as sources of direct experience 45, 46; subject at centre of, 88–9, 90; territorialisation 87; visibility of 89–90; *see also* performative mapping
Marres, Noortje 231–2
Martí, Alejandro 153
Mascolo, Cecilia 107
Massey, Doreen 98
materiality 10–13; of communication 10–11; new materialism 10
materiality: of communication practices 219; of concrete things 219
McChesney, Robert 126
McDonald, Barbara 143
McKelvey, Fenwick 218–19
McLuhan, Marshall 217
media archaeology 217
media ecologies 217
media geography 5
Mercado, Tomás Torres 153
Mercator projection 88
Merhmand, Elle 66
Metro México app 150
Mexico: apps industry 150; migration from 70; mobile and locative media in, 150–1; mobile network operators 150; privacy policy 14, 148, 151–2; *see also* Geolocalization Law; Transborder Immigrant Tool
Mi Policía app 151
Microsoft 126
Mitchell, J. W. T. 101–2
Mixi 6
mobile intimacy 32
mobile phones 5
mobility 7
Mol, Annemarie 110
Molinari, Susan 139
Monmonier, Mark 89
mood monitoring 199–201
MoodScope 197, 200–1
Murdock, Graham 178

Nagel, Till 77
Najarro, Jason 66
National Human Rights Commission (CNDH) (Mexico) 155
National Survey on Victims and Public Perception of Safety (Mexico) 154
Negri, Antonio 227
Netflix 109
new materialism, *see* materiality
New York: sitting conventions 59–60; study of public parks 52, 56–8; tracing experiences 59–62
Newman, Mark 113–14
Nieto, Enrique Peña 157
Nissenbaum, Helen 124–5
Nold, Christopher 90
non-participation in locative media: cost of 165
Noulas, Anastasios 54, 107

object-oriented ontology (triple O) 12
objects: force of 10–13
occluded culture 165
Okabe, Daisuke 23, 30, 32
Oliver, Julian 10, 77, 97–9, 227, 233; *Border Bumping* 77, 97–8; *Transparency Grenade* 227, 233
opt-out policies 165–7; personal cost 165–6
orientation practices, *see also* maps

Packer, Jeremy 10–11
Pandora 109
Papacharissi, Zizi 123
Parks, Lisa 69, 217
Partido Acción Nacional (PAN) 153
Peirce, Charles Sanders 86
Perec, Georges 53
performative mapping, 94–5, 100; performance of space 98–100
Perkins, Chris 87
Peters, John Durham 210
Pew Research Centre: State of News Media report 126–7; Internet and American Life Project 166
Pink, Sarah 27
Pinterest 123

Please Rob Me 131, 163, 168–9, 170
Poell, Thomas 178
points of interest 107
Polak, Esther 9, 90–1, 231–2; *Amsterdam REALtime* 90; *Souvenir* 91; *NomadicMILK* 91; *Spiral Drawing Sunrise* 231–2
political economy 13–14
politics of resistance: anthropocentric 224–5
Pontil, Massimiliano 107
portable networked media technologies 5
privacy: changing conceptions of 121–33, 165–7; child audience 123; contextual integrity 125; in convergent media industries 122–5; individual agency 123; law in Australia 140–2, 144, 145; locative media and 13–14; media corporations and 121; paradox 166; policy implications 122; policy intervention 124, 129–30, 132–3, 164–5; *see also* do not track (DNT); Geolocalization Law (Mexico); Google Glass; opt-out policies; promotional culture; user data
Proboscis collective 100; *Urban Tapestries* 100; *Mimoa* 100; *Historypin* 100
promotional culture 126–30; digital advertising 126–7; hyper-targeting technologies 126; regulatory scrutiny 129–30; *see also* cookies
Propuesta Cívica 156
proximity 111–12
public spaces: particularity of 55; reading through geotagged data 52–64; social media and 8–9; social media signals in 63; study of 52–6; spatial reading 56–8; temporal reading 58–9; tracing experiences in 59–62
public–private boundaries: locative media and 163

QQ 6

Rainert, Alex 179
Raley, Rita 75–6
Regan, Priscilla 125
Relph, Ed 3
Renren 6
representational paradigm 9
representations of everyday reality 7–10
RFID technology 6, 201, 226
Richardson, Ingrid 164, 165
Richardson, Megan 143
Ronnenberg, Philip, 10, 227, 233–4; *OpenPositioningSystem*, 233–4; *Post-Cyberwar* 227, 233–4; *The Sewer Cloud* 234; *Teletext Social Network* 234
Rousseff, Dilma 151
Rueb, Terri 96; *Elsewhere/Anderswo* 96
RunKeeper 47
Russell, Ben 225; *Heapmap Manifesto* 225

Salazar, José Adán Rubí 156
SAP 12, 208, 209; supply chain software 213–15
Scellato, Salvatore 107
Schiller, Dan 177
Schimanski, Johan 72
Se me antoja app 150
Seamon, David 40
sel-ca 25
self-portraiture on camera phones, *see* camera phones
Selvadurai, Naveen 179
Sennett, Richard 114
Shannon–Weaver theory 99
Sharma, Sarah 84–5
Sheller, Mimi 217, 219
Shepard, Mark 96; *Hertzian Rain* 96
Sicilia, Javier 153
Simcoe, Luke 218–19
Simmel, Georg 44
Simondon, Gilbert 100
Skees, Chelsa 181
Sleight, Peter 109
Slim, Carlos 150
smartphone 5, 6, 23, 121, 148; 4G network 6; 5G network 6; as hand-held drone 197, 198–9; limiting ads 129–30; as source of locational information 44; *see also* Android platforms; camera phone; drones; iPhone; locative surveillance
Snapchat 25, 26
Snowden, Edward 151, 158, 193
Sobchack, Vivian 87
social networking apps, mobile 4

soft infrastructure 215–17
Souriau, Étienne 231
spatial experience: embodied engagement 85; emphasis in mobile culture 83; impact of mobile maps 39–50; of locative media 100–2; mediation by ICTS 94; performance of space 94–103; space as passive surface 99; space as social 85; *see also* maps
spreadable media 167–8
Stalbaum, Brett 66, 72, 73, 76
Standard Operating Procedure (SOP) 212
Stole, Inger 126
Strabo 2–3
Strathern, Marilyn 229–30
Streetmuseum project 86–7, 88, 89
Studio Moniker 100; *What If Google Maps Went Live* 100
Sun Microsystems 193
surveillance: microtechnologies of 210; *see also* locative surveillance
Sutton-Smith, Brian 29

Tarkka, Minna 3
Tavera, Pilar 156
Taxibeat app 150
technology as magical 163–4
Thrift, Nigel 9, 101
Tiessen, Matthew 218–19
topology 110; nearest-neighbour network 111–12
Transborder Immigrant Tool 8, 66–8, 69, 71–5, 77, 78; historical framing 75; interface 73
transduction 100
transparency 162–71
Trenti, Fernando Jorge Castro 153
Tuters, Marc 228
TwitPic 170
Twitter 2, 68, 121, 123, 150, 168, 180
two-fold directedness 87

UniNet, 150, 156
US Air Force 68
US congressional privacy caucus 139
US Federal Trade Commission (USFCT) 123, 129
user data: data mining 196, 202; marketing of 193; metadata 194; protection of 125; stockpiling of 121; transparency regarding 162–71; uses for 193–4; *see also* locative surveillance; logistical media
user-generated content (UGC) 177
Ushahidi 1, 148

van Couvering, Elizabeth 185
Van Dijck, José 178
Vargas, Jorge A 151–2
Verizon 183
vibrant materialism 202
Villaneueva, Raúl Plasencia 155
Villi, Mikki 27
Vimeo 123
Vine 187
Virilio, Paul 227
Virtual Hiker 73
visuality: multisensoriality of images 27; shift from networked to emplaced 26–7

Wallace, Isabel Miranda de 153
Ward, Matt 225
Warden, Pete 169
Waters, Nigel 142
Watson, Jeff 183
Waze 187
Webber, Richard 109
WeChat 1
Weibo 2
Weiss, Helmut 214–15
Westlake, E. J. 166–7
Whyte, William 53
Wi-Fi 3, 5, 97
Wiley, Stephen Crofts 10–11
Wilken, Rowan 68, 131, 168
Wilson, Jake 212
Winseck, Dwayne 178
Wodiczko, Krzysztof 67–8, 77: *Alien Staff* 67–8, 77
World Game Project 225
Wray, Stefan 72

Yahoo 122, 126
Yelp 53, 55, 56, 88, 148
YFrog 170
YouTube 6, 121, 123

Zapatista FloodNet 72–3
Zarsky, Tal 204
Zhang, Amy X. 54
Zittrain, Jonathan 164
Zook, Matthew 11